Lecture Notes in Computer Science 12753

More information about this subseries at http://www.springer.com/series/7409

Gang Pan · Hui Lin · Xiaofeng Meng ·
Yunjun Gao · Yong Li ·
Qingfeng Guan · Zhiming Ding (Eds.)

Spatial Data and Intelligence

Second International Conference, SpatialDI 2021
Hangzhou, China, April 22–24, 2021
Proceedings

Springer

Editors
Gang Pan
Zhejiang University
Hangzhou, China

Xiaofeng Meng
Renmin University of China
Beijing, China

Yong Li
Tsinghua University
Beijing, China

Zhiming Ding
Chinese Academy of Sciences
Beijing, China

Hui Lin
Jiangxi Normal University
Nanchang City, Jiangxi, China

Yunjun Gao
Zhejiang University
Hangzhou, China

Qingfeng Guan
China University of Geosciences
Wuhan, China

ISSN 0302-9743   ISSN 1611-3349   (electronic)
Lecture Notes in Computer Science
ISBN 978-3-030-85461-4   ISBN 978-3-030-85462-1   (eBook)
https://doi.org/10.1007/978-3-030-85462-1

LNCS Sublibrary: SL3 – Information Systems and Applications, incl. Internet/Web, and HCI

This Springer imprint is published by the registered company Springer Nature Switzerland AG
The registered company address is: Gewerbestrasse 11, 6330 Cham, Switzerland

Preface

This volume contains the papers from the ACM Spatial Data Intelligence China Conference (SpatialDI 2021). The first edition of SpatialDI was held online due to the impact of COVID-19. The 2nd edition (SpatialDI 2021) was held in Zhejiang Hotel, Hangzhou, China, 22–24, April 2021.

SpatialDI 2021 mainly aimed to address the opportunities and challenges brought about by the convergence of Computer Science, GIScience, AI, and beyond. The main topics of the conference were Spatial Machine Learning and Artificial Intelligence, Spatial Data Acquisition and Positioning, High-Performance Computing of Large-scale Spatial Data, Mobile Data Management and Analysis, Geographic Information Retrieval, Spatial Semantic Analytics, Autonomous Transportation and High-precision Maps, Urban Analytics and Mobility, Spatial-temporal Visualization and Visual Analytics, Location-Based Services and Privacy Issues, Geo-social Network Analytics, Geo-computation for Social Science.

We received 72 submitted contributions for SpatialDI 2021. All of the submitted papers were assigned to three members of the Program Committee (PC) for peer review. All reviews were checked and discussed by the PC chairs and additional reviews or meta-reviews were elicited if necessary. Finally, we accepted 30 papers for SpatialDI 2021, with an acceptance rate of 41.67%.

In addition to regular papers, the conference invited Academician Huadong Guo (a member of the Aerospace Information Research Institute, Chinese Academy of Sciences), Academician Renzhong Guo (a member of the Chinese Academy of Engineering, Dean of the Research Institute for Smart Cities, Shenzhen University), Prof. Hui Lin (a member of the International Eurasian Academy of Sciences, Dean of the School of Geography and Environment, Jiangxi Normal University), Prof. Yunhao Liu (Dean of the Global Innovation Exchange, Tsinghua University, ACM/IEEE Fellow), Prof. Christian S. Jensen (Aalborg University, Fellow of the European Academy of Engineering, ACM/IEEE Fellow), and Prof. Jianwei Yin (Associate Dean of the College of Computer Science and Technology, Zhejiang University) to give the keynotes. The event was organized into 8 forums: Big Earth Data and the United Nations SDG; Spatial Intelligent Computing; Social Geographical Computing; Spatial-temporal Data Management; Blockchain and Data Services; Urban Computing; Location-based Services and Smart Travel; and Rising Star Award and the Doctoral Dissertation Award Special Session.

The proceedings editors wish to thank our keynote and invited speakers and all the reviewers for their contributions. We also thank Springer for their trust and for publishing the proceedings of SpatialDI 2021.

Gang Pan
Hui Lin
Xiaofeng Meng
Yunjun Gao
Yong Li
Qingfeng Guan
Zhiming Ding

Organization

Consultant Committee

Huadong Guo	Aerospace Information Research Institute, Chinese Academy of Sciences, China
Chenghu Zhou	Institute of Geographic Sciences and Natural Resources Research, Chinese Academy of Sciences, China
Jianya Gong	Wuhan University, China
Qingquan Li	Shenzhen University, China
Xiaofang Zhou	Hong Kong University of Science and Technology, Hong Kong, China

General Conference Chairs

Gang Pan	Zhejiang University, China
Hui Lin	Jiangxi Normal University, China
Xiaofeng Meng	Renmin University of China, China

Program Committee Chairs

Yunjun Gao	Zhejiang University, China
Yong Li	Tsinghua University, China
Qingfeng Guan	China University of Geosciences, Wuhan, China

Publication Chair

Zhiming Ding	Institute of Software, Chinese Academy of Sciences, China

Industrial Chairs

Feifei Li	Alibaba, China
Hua Chai	DiDi, China

Sponsorship Committee Chair

Danhuai Guo	Computer Network Information Center, Chinese Academy of Sciences, China

Local Arrangements Chairs

Xiaoye Miao	Zhejiang University, China
Sha Zhao	Zhejiang University, China

Publicity Chairs

Lu Chen	Zhejiang University, China
Longbiao Chen	Xiamen University, China
Xiao Pan	Shijiazhuang Tiedao University, China

Workshop Chairs

Jianliang	Hong Kong Baptist University, Hong Kong, China
Weiwei Sun	Fudan University, China
Bin Yao	Shanghai Jiao Tong University, China
Yang Le	Shenzhen University, China
Feng Zhang	Zhejiang University, China
Jianqiu Xu	Nanjing University of Aeronautics and Astronautics, China
Xiaoping Du	Aerospace Information Research Institute, CAS, China
Bolong Zheng	Huazhong University of Science and Technology, China
Hua Lu	Roskilde Universitet, Denmark
Jilin Hu	Aalborg University, Denmark
Yajin Zhou	Zhejiang University, China
Zhongchang Sun	Aerospace Information Research Institute, CAS, China

Program Committee

Lu Chen	Zhejiang University, China
Longbiao Chen	Xiamen University, China
Peng Cheng	East China Normal University, China
Yixiang Fang	The University of New South Wales, Australia
Danhuai Guo	Computer Network Information Center, CAS, China
Jilin Hu	Aalborg University, Denmark
Hao Huang	Wuhan University, China
Yang Yue	Shenzhen University, China
Huan Li	Aalborg University, Denmark
Bohan Li	Nanjing University of Aeronautics and Astronautics, China
Ronghua Li	Beijing Institute of Technology, China
An Liu	Soochow University, China
Hua Lu	Roskilde University, Denmark
Xiaoye Miao	Zhejiang University, China
Xiao Pan	Shijiazhuang Tiedao University, China

Contents

City Analysis

Traffic Management

A Deep Urban Hotspots Prediction Framework with Modeling Geography-Semantic Dynamics

Hengyu Sha[✉], Guangyin Jin, Guangquan Cheng, Jincai Huang, and Kuihua Huang

College of System Engineering, National University of Defense Technology, Changsha, China

Abstract. The research focus of this paper is based on the adaptive learning method for spatio-temporal situation prediction, focusing on solving some common and difficult problems in spatio-temporal prediction and some urgent needs in practical applications in the scenario of network data or graph structure data modeling. Problems, such as capturing the correlation between space and time in complex dynamics, extracting the internal correlation of non-neighborhood nodes in a deeper non-Euclidean space, improving the accuracy of the spatiotemporal situation prediction model, and making the model possess Difficulties such as adaptive learning features. Carry out research on spatiotemporal data preprocessing problems, potential correlation mining problems between geographic space and semantic space, spatiotemporal situation prediction problems, feature fusion problems, etc., and develop graph-based spatiotemporal prediction models with good universality and migration, and evaluation indicators It is compared with the traditional deep learning-based spatio-temporal prediction model, which proves that the graph-based spatio-temporal prediction model will achieve higher prediction accuracy of spatio-temporal signals and has better interpretability for many non-Euclidean spatial relationships in the real world.

Keywords: Spatiotemporal prediction · Smart city · Semantic modeling · Predictive recurrent neural network · Graph convolutional neural network

1 Introduction

According to the "Urban Development Report" of the World Bank [1], the process of urbanization is getting faster and faster. The rapid urbanization process has brought huge amounts of data to be processed. Based on urban big data, we can discover some meaningful patterns in many fields such as traffic management, environmental monitoring, and urban safety to guide urban planning and decision-making. This type of technology is the foundation of future smart city construction. We define it as "urban computing", and its concept was proposed by Zheng Yu and others in 2014 [2]. In order to understand the complex spatio-temporal dynamics in the urban system, spatio-temporal prediction is one of the most basic and important links in urban calculations [3].

The biggest challenge in predicting the distribution of urban spatio-temporal hotspots is to capture the correlation between space and time from the complex dynamics. The successful application of convolutional neural network (CNN) [4] and recurrent neural

© Springer Nature Switzerland AG 2021
G. Pan et al. (Eds.): SpatialDI 2021, LNCS 12753, pp. 3–14, 2021.
https://doi.org/10.1007/978-3-030-85462-1_1

network (RNN) in image recognition and time series processing, respectively, has promoted the development of deep learning in spatio-temporal prediction. In addition, with the advent of the era of big data, rich urban perception data provides a solid foundation for deep learning. Therefore, the spatio-temporal prediction method based on deep learning has been widely used in recent years. In the existing work, the most common data preprocessing method is to divide the urban area into small m*n grids according to the latitude and longitude. The hotspot statistics in each grid can be regarded as the pixel value of the spatial heat map (SHM), while the temporal heat map is the initial input of the CNN-based model [5]. This method is intuitive and can be easily captured by the CNN model. But the previous works still have certain limitations. (a) The distribution of hot spots in geographic grid space only reflects the spatial correlation of Euclidean space [6], while ignoring some deeper internal correlations of non-Euclidean space. We know that there are many different functional areas such as residential, commercial, and industrial in urban areas [7]. There are some implicit long-term correlations between these areas, even if they are not geographically adjacent. (b) In order to improve the accuracy of the prediction model, the traditional method is to integrate more additional relevant information. However, some additional data is difficult to obtain, and auxiliary information can only work in specific areas. This means that the portability and scalability of the model is limited [8].

In order to solve the above problems, we proposed the Geographic Semantic Integrated Neural Network (GSEN), which integrates geographic information and self-semantic information without any external auxiliary information. From a geographic perspective, we first process spatio-temporal data into spatial heat maps (SHMs); from a semantic perspective, we use spatial correlation coefficients to model spatial self-semantic maps (SSGs). Based on Predictive Recurrent Neural Network (PredRNN), this paper proposes a graph convolution predictive recurrent neural network (GC-PredRNN) model for processing SSG. Then, predict the PredRNN and GC-PredRNN to capture the spatio-temporal correlation from the SHM and SSG sequences, respectively. Subsequently, the geographical and semantic-based predictions are fused by the integrated convolutional layer. This model establishes the potential correlation between geographic space and semantic space, and the limitations of traditional deep learning models can be overcome by the addition of SSG. In summary, we have made the following main contributions in this article:

(1) As far as we know, this is the first time that a general geographic semantic information fusion framework has been proposed to solve the spatiotemporal prediction task of multi-city hotspots.
(2) Our proposed model GSEN is an efficient calculation framework without any other additional data. Compared with other frameworks that rely on data, this improves its portability and scalability.
(3) We evaluated our method on three real city data sets. The results show that, compared with the traditional method, this method almost reduces the errors of all evaluation indicators, indicating that the GSEN model has superiority and universality in the temporal and spatial prediction of urban hot spots.

2 Related Work

2.1 Prediction of Temporal and Spatial Situation Based on Mathematical Modeling

The self-exciting point process (self-exciting point process) is one of the most commonly used mathematical models for temporal and spatial prediction, and the modeling of events depends on its past history [9]. The model has proven useful in many fields: seismological models of earthquakes and aftershocks, prediction models of crime dynamics, epidemiological predictions of disease incidence, and many other fields. In each field, the temporal and spatial distribution of events has scientific and practical significance. It can predict new events and enhance the understanding of the process of event generation. The self-excitation point process was first applied to earthquake prediction models. Earthquake and aftershock sequences show rich behaviors, such as temporal and spatial clustering, complex spatial correlations, and gradual changes in overall seismic activity. The self-excitation point process can directly capture the triggering behavior of spatiotemporal aftershocks, and can be compatible with temporal trends and spatial inhomogeneities. The popular type of aftershock sequence (ETAS) model [10] has been widely used in earthquake sequence prediction in Japan, California and other regions after decades of development and expansion. After developing the ETAS model, Mohler et al. made an analogy between the aftershock model and crime and developed a spatiotemporal point process model suitable for crime prediction [11]. On the basis of earthquake and crime prediction models, the spatiotemporal self-excitation point process has been gradually extended to epidemic spread prediction [12], social network event mining [13] and other fields. When the spatiotemporal point process can be divided into event clusters triggered by common causes, the self-excited model is a powerful tool for simulating the dynamics of the spatiotemporal process. The advantage of this type of model is that it has good mathematical interpretation, but they still have many unsolved problems. The biggest problem is extremely low generalization. Different mathematical models must be established for the same type of data from different sources (for example, crime data from different cities), and the application of self-excited models to network data is still in its infancy.

2.2 Spatio-Temporal Prediction Based on Statistical Learning

Compared with mathematical modeling methods, the spatiotemporal data prediction method based on statistical learning is more convenient and concise. It does not require multiple modeling for uniform types of tasks. Support vector machines, decision trees and other methods are also interpretable, so it has been developed over a long period of time. As early as 2010, a support vector machine multi-class classifier based on a binary tree structure was proposed to identify multiple activities from the system framework of the video. This model combines the advantages of the support vector machine and the decision tree model. It is statistical learning in humans. The first attempt in the field of spatiotemporal behavior prediction [14]. Since then, spatio-temporal prediction methods based on statistical learning models have been extended to various fields. For example, in the field of weather forecasting and environmental detection, Xu Quanjun, Xie Zhimin and others have proposed a decision tree-based short-term thunderstorm forecast model. The decision tree model integrates temporal and spatial characteristics to predict the occurrence of thunderstorms in the next 3–4 h. Above 90% [15]. On this basis, Yu Zhan et al. proposed a PM2.5 spatiotemporal distribution prediction model GW-GBM based on an improved gradient ascent tree, which solved the spatial non-stationarity of PM 2.5 dependence on environmental conditions, thereby greatly improving improved predictive performance [16]. In the field of smart transportation, Yingshan Zhong et al. proposed a demand forecasting model for shared bicycle stations based on random forest and spatio-temporal clustering. The model studied the influence of time factors, meteorological factors and related stations on demand, and applied hierarchical application. Clustering conducted a spatio-temporal analysis of the sites, combined with a logarithm-optimized random forest as a predictor to accurately predict the demand for urban shared bicycle sites [17]. Haowei et al. also proposed a short-term traffic flow prediction model based on support vector regression, which has relatively high prediction accuracy [18]. The spatio-temporal prediction method based on the statistical learning model has been very popularized and applied, and has achieved good results in many fields. But the premise is that it must be supported by good feature engineering. If there is no good manual selection of features, it is difficult to achieve high-level performance. This is the biggest limitation of the spatiotemporal data mining method based on the statistical learning model.

2.3 Spatio-Temporal Prediction of Deep Learning Based on Grid Representation

With the advent of the era of big data and the classic deep neural network models CNN and RNN have made major breakthroughs in image recognition, sequence learning and other fields, the spatiotemporal data prediction method based on deep neural networks has gradually become one of the most mainstream methods. Xingjian et al. proposed the Conv-LSTM model in 2015 [19], which is the first combination of CNN and RNN series models and used in the field of weather forecasting, proving that this type of hybrid model is significantly better than other traditional methods. On this basis, more and more hybrid deep neural network models have begun to emerge and are widely used in the field of spatio-temporal prediction. The first step of data processing of this spatio-temporal prediction model based on deep neural networks is often to divide the spatial data into uniform grids and use them as the pixel representation input of the CNN model, so it is also called depth based on grid representation. Learn about spatio-temporal prediction. Among them, the most relevant research work is in the application of urban computing, smart transportation and other fields. On the basis of the most popular residual network (ResNet) in the field of computer vision (ResNet) [20], Junbo Zhang et al. proposed a spatiotemporal residual network model (ST-ResNet), which uses residual neural networks to learn spatial distributions of different time scales and combining features to achieve accurate predictions for the flow of people and traffic in the city [21].

Similarly, Bao Wang et al. improved the ST-ResNet model and transferred it to the real-time prediction scenario of urban crime situation, and verified its effectiveness on the real crime record data set in Los Angeles [22]. With the introduction of sequence-to-sequence learning seq2seq [23] and some models based on attention mechanisms, many works have begun to explore the long-term dependence of multi-step spatiotemporal prediction methods and time series. In the work of predicting traffic flow speed based on deep sequence learning proposed by Binbing Liao et al., a multi-step prediction model based on the seq2seq model was developed [24]. Zheng Yu and others also proposed DeepCrime, a hierarchical attention loop neural network model, which not only proves that it has a good predictive effect on urban crime dynamics, but also visualizes the attention weights of different crime types and different time steps to analyze specific crimes. The degree of influence of the type on the future development trend at a specific time step. Some of the most popular deep neural network models such as Generative Adversarial Networks (GAN) have also been combined and used in the field of spatiotemporal prediction and mining. The method based on the GAN model was first applied to video prediction. The work of Michael et al. combined the GAN model for the first time to improve the realism of the predicted time series image [25]. D-GAN [26] proposed by Divya Saxena et al. adopts a GAN-based structure, and simultaneously learns the generation of data and variational inference, and simulates the changes in the amount of spatio-temporal data over time. Compared with methods based on statistical learning, the biggest advantage of spatio-temporal data mining methods based on deep neural networks is that they do not require manual selection of features. The powerful automatic extraction capabilities of deep neural networks save a lot of feature engineering costs, but deep neural networks the method's interpretability, robustness, and data dependence have yet to be resolved. In addition, the input of spatiotemporal data mining methods based on deep neural networks is mostly represented by pixel-level features. However, there are still many difficulties in the processing of networked data or non-Euclidean spatial structure related data in some specific tasks.

2.4 Spatio-Temporal Prediction Based on Dynamic Graph Learning

In the real world, such as e-commerce data, social data, transportation network data, etc., more rigorous non-European spatial representations can be obtained through graph structures. Therefore, the mining of graph data in the real world has been the work for the past five years Hot spot. Obtaining an effective low-dimensional representation of graph structure is the core task of all graph data mining tasks. The earliest are some graph representation learning methods based on factorization [27]. Since 2014, many classical methods of graph representation learning based on random walks have been proposed, such as DeepWalk [28] and node2vec [29] which are commonly used classical methods so far. In 2016, the work of kipf and Welling brought graph data mining to a new climax. They proposed the semi-supervised classification of graph nodes using the graph neural network (GCN) model [30]. Since then, representation learning, node classification, and link prediction based on the graph neural network model began to blow out. Typical examples include the variational graph autoencoder (VGAE) model proposed by kipf and welling, which efficiently obtains low-dimensional representation of graph data and performs link prediction [31]. The GraphSAG model proposed by Hamilton et al. makes distributed computing of large-scale graph data possible [32]. Guillem et al. also proposed the Graph Attention Network (GAT), which uses the attention mechanism to aggregate neighbor nodes to realize the adaptive allocation of different neighbor weights [33]. Although many remarkable results have been achieved in the field of graph data mining, most of the past work is based on static graph data scenarios. Many graph structure data in the real world change over time, but the spatiotemporal data mining method based on dynamic graph learning has only just started in the past two years. Based on the neural network model Conv-LSTM, Youngjoo et al. proposed a graph convolutional recurrent neural network (GCRN) model [34], which is similar in structure to Conv-LSTM, but replaced the CNN model with the GCN model, successfully transferring the spatio-temporal prediction model to Under the scene of dynamic graph data [35]. On this basis, more graph neural network models based on dynamic graphs have begun to be proposed, but so far, most of the research has focused on the field of smart transportation.

3   Main Research Content

3.1   Spatio-Temporal Data Modeling Method Based on Adaptive Semantic Graph

In the existing research on temporal and spatial situation prediction, the most common data preprocessing method is to divide the urban area into m*n small grids according to the latitude and longitude. The hot spot statistics in each grid can be regarded as the pixel value of the spatial heat map (SHM), and the time heat map is the initial input of the CNN-based model. However, this method is based on the standard Euclidean space and only considers the influence between neighbors, while ignoring the fact that most of the regional nodes in the reality are not The objective fact of Euclidean space. Without considering the deeper internal correlations of non-Euclidean nodes, it is more one-sided and cannot make a reasonable and accurate prediction of the temporal and spatial situation. In this research, we introduce the concept of adaptive semantic graph, and construct an adjacency matrix after gridding the original data, and then construct an adaptive semantic graph, using the concept of nodes and edges in the graph structure to represent the actual nodes The mutual influence breaks the limitation of the traditional Euclidean space, so that the spatio-temporal data input to the neural network is more comprehensive and reasonable.

Take the urban hotspot data set as an example. In order to avoid some data being too sparse and incomplete, we should adjust the space and time range of the original data. In order to obtain sufficient statistical data, we should first classify the events and then proceed according to different time spans. Statistics, for example, determine a week as a time period in the theft crime data collection, two hours as a time period in the online car-hailing data collection, and a day as a time period in the fire data collection. Then use the data set after preliminary processing to construct a spatial heat map. The method of constructing the spatial heat map is: Given the area grid index and the time slot index t, we define the element x_{ij}^t as a specific event $X_t = [x_{ij}^t]_{I*J}$. The number of occurrences in this space-time background (depending on the type of event, the selected time interval is also different), where i*j is a geographic indicator. Furthermore, grid processing is performed to effectively segment the target area. The following figure is an example. The target area is divided into a 10×10 grid, that is, it is subdivided into 100 sub-areas. This can grid the input geographic area to meet the input requirements of the convolutional neural network, but this only takes into account the proximity relationship in the geographic space. In fact, there are many non-geographic proximity relationships in urban hot spots (such as Railway stations and hospitals can affect areas that are geographically remote from the area where they are located, not only the situation of their neighborhoods), so we should find a way to incorporate this relationship into the model at the same time The scope of consideration to enhance the accuracy and rationality of the model. Input the spatial heat map into the prediction recurrent neural network, and input the self-semantic map into the prediction recurrent neural network, and finally get the respective output of the two neural networks, and then through the integration layer for feature fusion, a new feature matrix is generated for output. After the feature matrix is gridded, it is transformed into a spatial heat map to make people feel the prediction effect more intuitively (Fig. 1).

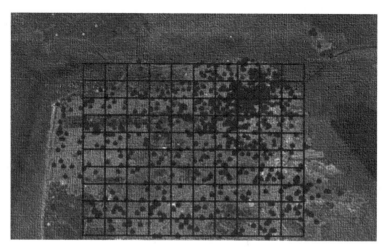

Fig. 1. Schematic diagram of 10 × 10 division processing of spatial heat map grid.

3.2 Mining the Spatio-Temporal Correlation of Situational Elements Fused with Semantic Information

In the past, deep learning-based spatiotemporal situation prediction methods have certain limitations. The traditional CNN model method based on time heat map training has the following two problems: (a) The hotspot distribution in geographic grid space only reflects traditional Euclidean The spatial correlation of space, while ignoring the internal correlation of some deeper non-Euclidean space nodes. We know that there are many nodes in urban areas. Even if they are not geographically adjacent, due to their respective functional differences, there are some implicit long-term correlations between these areas, which will affect each other's time and space situation. The forecast will have a certain impact. (b) In order to improve the accuracy of the prediction model, the traditional approach is to incorporate more additional relevant information into the model. However, some related data is difficult to obtain, and auxiliary information is only valid in a specific field, which means that the portability and scalability of the model is greatly restricted. Although some external information fusion improves the prediction performance of the model and can be associated with some real-world indicators, it sacrifices the transferability and scalability of the model, and the acquisition of external data and a larger computational burden on the model The performance brings a bottleneck.

In this study, we mine some potential high-level self-semantic information from the original data and fuse it with geospatial-based grid images to improve the performance of spatio-temporal prediction. It is expected that a new geographic semantic integration neural network based on deep learning methods is proposed, which superimposes the geographic prediction neural network and the semantic prediction neural network. The model includes the structure of Predictive Recurrent Neural Network (PredRNN), Graph Convolutional Prediction Recurrent Neural Network and integrated layer. It captures spatiotemporal dynamics from different angles and makes effective predictions, and then unearths some potential high-level dynamics in spatiotemporal data. Any additional prior knowledge is required, with adaptive learning characteristics.

Therefore, in order to achieve the above goals, we first start from a geographical point of view, and on the basis of processing related data sets into spatial heat maps (SHMs); then from a semantic point of view, use spatial correlation coefficients to model spatial self-semantic maps (SSGs).

The method of constructing the spatial self-semantic graph is: given the area grid point index (i, j) and the time statistics x_{ij}^t of each grid, take a grid area as the vertex and use the absolute Pearson correlation coefficient Construct an adjacency matrix. The feature matrix of the SSG is the statistics of the number of events that occur at each node based on the time span. The adjacency matrix of SSG is defined as:

$$G(V, E), \ r_{ij} = \begin{cases} p_{ij} = \left| \dfrac{\sum_{i=1}^{n}(x_i-\bar{x})(y_i-\bar{y})}{\sqrt{\sum_{i=1}^{n}(x_i-\bar{x})^2}\sqrt{\sum_{i=1}^{n}(y_i-\bar{y})^2}} \right| \\ 0 \quad if \ p_{ij} < 0.05 \end{cases}$$

$$A = \begin{bmatrix} r_{0,0} & r_{0,1} & \cdots & r_{0,9} \\ r_{1,0} & r_{1,1} & \cdots & r_{1,9} \\ \vdots & \vdots & \ddots & \vdots \\ r_{9,0} & r_{9,1} & \cdots & r_{9,9} \end{bmatrix} \tag{1}$$

Among them, the adjacency matrix is stable in a certain time interval, and the characteristic matrix is constantly changing with time, and the adjacency matrix is constructed based on the long-term event statistics in each small grid. Therefore, the graph can reflect the long-term semantic correlation between different regions. As we discussed earlier, we can go beyond the limitations of Euclidean spatial geographic information. In addition, the construction of this graph does not require any other auxiliary information, so we call it a self-semantic graph. Use graphs to model a variety of correlations between regions, including (1) neighborhood graphs, coding spatial proximity (2) regional functional similarity graphs, and coding similarities of interest points around the area (3) traffic connectivity Sex map, which encodes the connectivity between distant regions. Adaptive: Input past spatiotemporal data parameters, and the model can be transformed into the corresponding adjacency matrix to express the connectivity and strength of each node, that is, the degree of influence of the inter-regional situation, such as the influence of the train station on remote areas during the Spring Festival (Fig. 2).

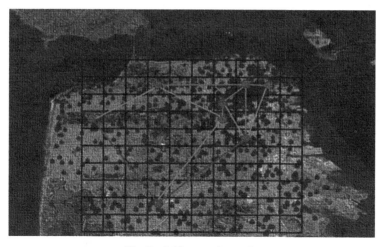

Fig. 2. Self-semantic graph.

3.3 Spatio-Temporal Situation Prediction Based on the Integration of Geographic Space and Semantic Space

Taking urban hotspot time and space situation prediction as an example, the most commonly used method for time and space situation prediction is to predict the distribution of urban hot spots on space and time scales. From a spatial perspective, the internal interactions between people and vehicles in different urban areas have an important impact on the distribution of urban hot spots. In terms of time, event hotspots in some urban areas exhibit cyclical or even seasonal fluctuations. Therefore, one of the major problems in spatio-temporal situation prediction is how to capture the correlation between space and time at the same time. In this study, predictive PredRNN and GC-PredRNN will be used to capture spatiotemporal correlations from spatiotemporal heat maps and adaptive semantic maps (Fig. 1). Subsequently, the geographical and semantic-based predictions will be fused by the integrated convolutional layer to establish the potential correlation between the geographical space and the semantic space. At the same time, in order to cope with the problems of gradient disappearance, gradient explosion, model overfitting, and computational resource consumption in deep learning, it is planned to use residual linking and attention methods to improve the performance of model learning deep features, so as to achieve more comprehensive and accurate Time and space situation forecast (Fig. 3).

3.4 The Basic Idea and Implementation Plan for Establishing the Model

Based on the Prediction Recurrent Neural Network (PredRNN) that combines temporal and spatial changes with LSTM, a graph convolution prediction recurrent neural network model for processing self-semantic graphs is proposed. Then, the prediction PredRNN and the graph convolution prediction recurrent neural network are respectively applied to capture the spatiotemporal correlation from the spatial heat map and the semantic map.

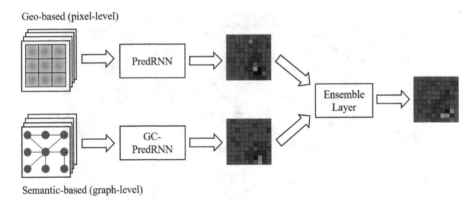

Fig. 3. Schematic diagram of the model.

Then we can apply the stacking method, which can obtain the final output by calculating the stacked output of multiple models. The motivation is that normal convolution and graph convolution pay attention to different categories of features (the PredRNN model focuses more on extracting geographic neighborhoods). The graph convolution prediction recurrent neural network model has a better effect on the extraction of node features in non-Euclidean space), through the fusion of non-linear operations, that is, the geographical and semantic-based predictions are integrated into the convolutional layer Fusion can achieve a more comprehensive perception of spatial characteristics. Use this model to establish the potential correlation between geographic space and semantic space, and through the addition of self-semantic maps, it can overcome the traditional deep learning model that is limited by traditional Euclidean space and cannot well dig into some deeper internal semantics. The limitations. As the number of neural network layers deepens, a series of problems are brought about, such as the disappearance of gradients, the explosion of gradients, the prone to overfitting of models, and the consumption of computing resources. As the number of network layers increases, network degradation will occur. The use of residual links can solve the above problems. By adjusting the convolutional neural network architecture, the channel attention method can make it easier for the model to learn deep features to improve the performance of the model.

After the effective training of a certain amount of data set, only the basic data sets of different time spans need to be parameterized and input into the model, and the model can generate the corresponding adjacency matrix by itself. The output of the final model is the feature matrix corresponding to the prediction time period, which can be converted into a spatiotemporal heat map to more intuitively show the hotspot distribution of the current prediction area, thereby forming a spatiotemporal situation prediction model with more accurate results and adaptive learning characteristics.

References

1. World Bank: Urban development report (2019)

2. Zheng, Y., Capra, L., Wolfson, O., Yang, H.: Urban computing: concepts, methodologies, and applications. ACM Trans. Intell. Syst. Technol. **5**(3), 1–55 (2014)
3. Atluri, G., Karpatne, A., Kumar, V.: Spatio-temporal data mining: a survey of problems and methods. ACM Comput. Surv. **51**(4), 83 (2018)
4. LeCun, Y., Bengio, Y.: Convolutional networks for images, speech, and time series. Handb. Brain Theory Neural Netw. **3361**(10), 1995 (1995)
5. Cheng, T., Wang, J.: Application of a dynamic recurrent neural network in spatio-temporal forecasting. In: Popovich, V.V., Schrenk, M., Korolenko, K.V. (eds.) Information Fusion and Geographic Information Systems, pp. 173–186. Springer Berlin Heidelberg, Berlin, Heidelberg (2007). https://doi.org/10.1007/978-3-540-37629-3_12
6. Wang, S., Cao, J., Yu, P.S.: Deep Learning for Spatio-Temporal Data Mining: A Survey, arXiv preprint arXiv:1906.04928 (2019)
7. Wang, Y., Long, M., Wang, J., Gao, Z., Philip, S.Y.: PredRNN: recurrent neural networks for predictive learning using spatiotemporal LSTMs. In: Advances in Neural Information Processing Systems, pp. 879–888 (2017)
8. Meyer, S., Elias, J., Höhle, M.: A space–time conditional intensity model for invasive meningococcal disease occurrence. Biometrics **68**(2), 607–616 (2012)
9. Mohler, G.O., Short, M.B., Brantingham, P.J., Schoenberg, F.P., Tita, G.E.: Self-exciting point process modeling of crime. J. Am. Stat. Assoc. **106**(493), 100–108 (2011)
10. Ogata, Y., Matsu'ura, R.S., Katsura, K.: Fast likelihood computation of epidemic type aftershock-sequence model. Geophys. Res. Lett. **20**(19), 2143–2146 (1993)
11. Williams, B.M., Hoel, L.A.: Modeling and forecasting vehicular traffic flow as a seasonal ARIMA process: theoretical basis and empirical results. J. Transport. Eng. **129**(6), 664–672 (2003)
12. Radziukynas, V., Klementavicius, A.: Short-term wind speed forecasting with ARIMA model. In: 2014 55th International Scientific Conference on Power and Electrical Engineering of Riga Technical University (RTUCON), pp. 145–149. IEEE (2014)
13. Pan, Y., Zhang, M., Chen, Z., Zhou, M., Zhang, Z.: An arima based model for forecasting the patient number of epidemic disease. In: 2016 13th International Conference on Service Systems and Service Management (ICSSSM), pp. 1–4. IEEE (2016)
14. Tchrakian, T.T., Basu, B., O'Mahony, M.: Real-time traffic flow forecasting using spectral analysis. IEEE Trans. Intell. Transport. Syst. **13**(2), 519–526 (2011)
15. Zarei, N., Ghayour, M.A., Hashemi, S.: Road traffic prediction using context-aware random forest based on volatility nature of traffic flows. In: Selamat, A., Nguyen, N.T., Haron, H. (eds.) ACIIDS 2013. LNCS (LNAI), vol. 7802, pp. 196–205. Springer, Heidelberg (2013). https://doi.org/10.1007/978-3-642-36546-1_21
16. Hong, W.-C.: Traffic flow forecasting by seasonal SVR with chaotic simulated annealing algorithm. Neurocomputing **74**(12–13), 2096–2107 (2011)
17. Yu, R., Yang, Y., Yang, L., Han, G., Move, O.A.: RAQ – A random forest approach for predicting air quality in urban sensing systems. Sensors **16**(1), 86 (2016)
18. Mcgovern, A., Supinie, T., Gagne, D., Troutman, N.P., Collier, M.W., Brown, R.A., Basara, J., Williams, J.: Understanding severe weather processes through spatiotemporal relational random forests. In: 2010 NASA Conference on Intelligent Data Understanding (to appear), Citeseer (2010)
19. Xingjian, S., Chen, Z., Wang, H., Yeung, D.-Y., Wong, W.-K., Woo, W.-c.: Convolutional LSTM network: a machine learning approach for precipitation nowcasting. In: Advances in Neural Information Processing Systems, pp. 802–810 (2015)
20. Wang, B., Yin, P., Bertozzi, A.L., Brantingham, P.J., Osher, S.J., Xin, J.: Deep Learning for Real-Time Crime Forecasting and its Ternarization, arXiv preprint arXiv:1711.08833 (2017)

21. Jin, G., Wang, Q., Zhao, X., Feng, Y., Cheng, Q., Huang, J.: Crime-GAN: a context-based sequence generative network for crime forecasting with adversarial loss. In: 2019 IEEE International Conference on Big Data (Big Data), pp. 1460–1469. IEEE (2019)

22. Yao, H., Tang, X., Wei, H., Zheng, G., Yu, Y., Li, Z.: Modeling spatial-temporal dynamics for traffic prediction, arXiv preprint arXiv:1803.01254 (2018)

23. Wen, C., et al.: A novel spatiotemporal convolutional long short-term neural network for air pollution prediction. Sci. Total Environ. **654**, 1091–1099 (2019)

24. Le, V.-D., Bui, T.-C., Cha, S.-K.: Spatiotemporal deep learning model for citywide air pollution interpolation and prediction. In: 2020 IEEE International Conference on Big Data and Smart Computing (BigComp), pp. 55–62. IEEE (2020)

25. Polson, N.G., Sokolov, V.O.: Deep learning for short-term traffic flow prediction. Transport. Res. C: Emerg. Technol. **79**, 1–17 (2017)

26. Wang, L., Geng, X., Ma, X., Liu, F., Yang, Q.: Crowd flow prediction by deep spatio-temporal transfer learning, arXiv preprint arXiv:1802.00386 (2018)

27. Seo, Y., Defferrard, M., Vandergheynst, P., Bresson, X.: Structured sequence modeling with graph convolutional recurrent networks. In: Cheng, L., Leung, A.C.S., Ozawa, S. (eds.) ICONIP 2018. LNCS, vol. 11301, pp. 362–373. Springer, Cham (2018). https://doi.org/10.1007/978-3-030-04167-0_33

28. Zhao, L., Song, Y., Deng, M., Li, H.: Temporal graph convolutional network for urban traffic flow prediction method, arXiv preprint arXiv:1811.05320 (2018)

29. Li, Y., Yu, R., Shahabi, C., Liu, Y.: Diffusion convolutional recurrent neural network: Data-driven traffic forecasting, arXiv preprint arXiv:1707.01926 (2017)

30. Zheng, C., Fan, X., Wang, C., Qi, J.: Gman: a graph multi-attention network for traffic prediction, arXiv preprint arXiv:1911.08415 (2019)

31. Geng, X., et al.: Spatiotemporal multi-graph convolution network for ride-hailing demand forecasting. In: 2019 AAAI Conference on Artificial Intelligence (AAAI 2019) (2019)

32. Wang, Q., Jin, G., Zhao, X., Feng, Y., Huang, J.: CSAN: a neural network benchmark model for crime forecasting in spatio-temporal scale. Knowl.-Based Syst. **189**, 105120 (2020)

33. Zhang, J., Zheng, Y., Qi, D.: Deep spatio-temporal residual networks for citywide crowd flows prediction. In: Thirty-First AAAI Conference on Artificial Intelligence (2017)

34. Yao, H., et al.: Deep multi-view spatial-temporal network for taxi demand prediction. In: Thirty-Second AAAI Conference on Artificial Intelligence (2018)

35. Wang, Y., Yin, H., Chen, H., Wo, T., Xu, J., Zheng, K.: Origin-destination matrix prediction via graph convolution: a new perspective of passenger demand modeling. In: Proceedings of the 25th ACM SIGKDD International Conference on Knowledge Discovery & Data Mining, pp. 1227–1235 (2019)

A Dynamic Traffic Community Prediction Model Based on Hierarchical Graph Attention Network

Lutong Li[1], Mengmeng Chang[1], Zhiming Ding[2(✉)], Zunhao Liu[1], and Nannan Jia[1]

[1] Beijing University of Technology, Beijing, China
{lilutong,changmengmeng,liuzunhao,jianannan}@bjut.edu.cn
[2] Institute of Software, Chinese Academy of Sciences, Beijing, China
zhiming@iscas.ac.cn

Abstract. The time-varying property of traffic networks has brought a problem of modeling large-scale dynamic networks. Based on the real-time traffic sensing data of the road network, community division and prediction can effectively reduce the complexity of local management in urban regions. However, traffic-based communities have complex topology and real-time dynamic features, and traditional community division and topology prediction cannot effectively be applied to this structure. Therefore, we propose a dynamic traffic community prediction model based on hierarchical graph attention network. It uses the hierarchical features fusion with spatiotemporal convolution and the ADGCN proposed in this paper to compose a hierarchical graph attention architecture. In which, each layer component coordinates to perform different features extraction for capturing traffic community of road network in different time periods respectively. Finally, the output features of each layer are combined to represent the dynamically divided regions in the traffic network. The effectiveness of the model was verified in experiments on the Xi'an urban traffic dataset.

Keywords: Traffic community division · Traffic prediction · Feature fusion · Graph convolution

1 Introduction

With the total number of vehicles in modern cities growing year by year, the traffic environment has become complex and changeable, which also increases the difficulty of traffic management. Dividing entire road network into subnet community according to the dynamic traffic information and predicting the divided community is one of the most effective methods for managing large road traffic network.

In the current research, using complex network theory to study the complexity of traffic network is one of the research hotspots [1, 2]. The current research on traffic network complexity mainly focuses on network modeling and community division. Related research shows that the traffic information of road network nodes has temporal and spatial correlation, and shows regional differences in the road traffic network [3]. According to the correlation of dynamic road traffic network, the traffic community

© Springer Nature Switzerland AG 2021
G. Pan et al. (Eds.): SpatialDI 2021, LNCS 12753, pp. 15–26, 2021.
https://doi.org/10.1007/978-3-030-85462-1_2

prediction can provide decision basis for urban traffic macro-management and improve the efficiency of urban traffic system.

The prediction of traffic community can be defined as the prediction question of t spatial-temporal sequence information. When performing spatial-temporal sequence prediction, not only the time continuity and periodicity, but also the spatial correlation of the road network nodes needs to be considered. In addition, the prediction of traffic community also needs to predict the topological structure of the divided regions. The traditional method based on statistical principle and machine learning cannot extract neighboring nodes dependency and regional features, so that can't effectively predict the topology of the divided regions.

In this paper, we adopt the method of deep learning to propose a dynamic traffic community prediction model based on hierarchical graph attention network. It uses three components for model architecture and each component builds a layer of graph attention structure through the attention based spatial-temporal graph convolutional network (ASTGCN [4]) and the attention based double-layer graph convolutional network ADGCN proposed in this paper. In which, the ASTGCN captures the spatial-temporal features of the traffic network through spatial-temporal convolution to predict the traffic information. The ADGCN calculates the classification probability matrix of road network nodes through the traffic information to obtain the topological structure information of traffic community. The three components capture the spatial-temporal features of recent, daily and weekly periods respectively, and integrate the output features to represent the dynamically divided regions in the traffic network. In the experiment of the Xi'an urban traffic dataset, we verified the effectiveness of the proposed dynamic traffic community prediction model based on hierarchical graph attention network through the comparison with other methods.

The main contributions of the paper are as follows:

1) The graph convolution network is used to solve the problem of road network community division for the first time, and the attention mechanism is integrated into the double-layer graph convolutional network to construct the ADGCN model, which effectively improves the accuracy of community division.

2) To solve the problem of predicting the topology structure of the dynamic traffic community, we use multiple components to compose a hierarchical graph attention network and construct the basic architecture of each component by fusing the AST-GCN and the ADGCN. Finally, we obtain the divided traffic community information through multi-component fusion.

3) Experiments of the Xi'an urban traffic dataset verify that the ADGCN is superior to the current community division methods in terms of Accuracy (AC) and Normalized Mutual Information (NMI), and also prove the effectiveness of the prediction model based on hierarchical graph attention network in the prediction of dynamic traffic community.

The remaining part of this paper is organized as follows. Section 2 reviews the related research. Section 3 introduces the definition of related problems. Section 4 describes in detail the hierarchical graph attention network. Section 5 presents the experimental evaluation results. Section 6 summarizes our conclusion.

2 Related Work

In recent years, scholars have carried out lots of research on the community division of dynamic traffic network. For example, Ramezani et al. established the aggregation model of the entire transportation network through the Macro Fundamental Diagram (MFD) [5], and used kinetic method to obtain the subnet structure of transportation network according to the difference of the road traffic flow. An et al. [6] used network nodes connectivity to determine the number of divided regions, and proposed a four-step network division method to achieve urban road network regional division.

With the rapid development of deep learning, graph convolutional network (GCN) is widely used in various fields, including community division of complex networks [7]. It uses double-layers convolution networks to derive a network embedding, and applies the softmax function to classify nodes into different categories. For example, Jin et al. integrate Markov random fields into graph convolutional neural networks to construct the MFCasGCN model [8], and proved the validity of the model in multiple real-world datasets.

The prediction of dynamic traffic community can be regarded as the prediction problem of the spatial-temporal sequence information. Williams et al. proposed the Autoregressive Integral Dividing Average Method (ARIMA) based on statistical principles [9], but this method is difficult to capture spatial correlation, and to dig out the nonlinearities of spatial-temporal pattern. Jeong et al. [10] used a machine learning based support vector regression method to capture non-linear relationships in spatial-temporal network nodes, but it is not effective in long-term prediction. With the rapid development of deep learning, Guo et al. [4] proposed a graph convolution model that integrates spatial-temporal attention and convolve a larger range of nodes correlation and achieve better predictions than current methods.

In addition, the prediction of dynamic traffic community also needs to predict regional information. Zhang et al. used rasterization to represent of urban road network [11], and proposed an ST-ResNet to capture the temporal and spatial dependence of different grids to achieve gird flow prediction. Zhou et al. used knowledge graphs to regionalize the road network, and proposed the KG-ST-CNN model to extract the spatial and temporal features to realize traffic congestion prediction in different urban region. Current methods are all based on static region of road network to predict traffic information, cannot fully represent the traffic status of all sections within the region. Therefore, using static region to present and predict regional traffic information has limitations.

3 Problem Definition

Definition 1: Dynamic Spatial Road Network. The spatial road network can be represented by a directed graph $G = (V, E, A)$, where V represents the set of nodes, E represents the set of edges, and A represents the adjacency matrix of the road network. The dynamic spatial road network can be defined as snapshots of the spatial road network based on time series: $\overline{G} = \{G_1 \dots G_t \dots G_L\}$, where G_t represents the spatial road network at time t. The representation of $\overline{G}$ is shown in Fig. 1(b).

Definition 2: Dynamic Community Road Network. According to the difference of traffic information in different regions of spatial road network G, it can be divided into multiple sub-network community. The community road network can be expressed as: $\ddot{G} = \{C_1 \ldots C_i \ldots G_K\}$, where C_i represents the i-th traffic community of the community road network $\ddot{G}$. And the dynamic community road network can be defined as snapshots of the community road network based on time series: $\ddot{\overline{G}} = \{\ddot{G}_1 \ldots \ddot{G}_t \ldots \ddot{G}_L\}$, where $\ddot{G}_t$ represents the community road network at time t. The representation of $\ddot{\overline{G}}$ is shown in Fig. 1(c).

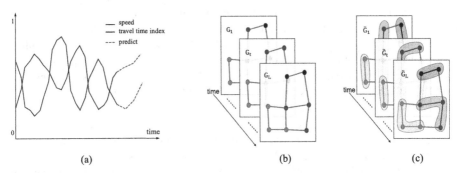

Fig. 1. (a) Two measurements are detected on a road network sensor node and predicted. Here, all measurements are normalized to [0,1]. (b) The dynamic spatial road network $\overline{G}$. (c) The dynamic community road network $\ddot{\overline{G}}$.

Definition 3: Dynamic Traffic Community Prediction. The prediction of the dynamic traffic community can be divided into the problem of traffic prediction and traffic community division. In which, traffic prediction can be regarded as the process of predicting future time series traffic information according to the collected data $X \in R^{N*F*L}$ of dynamic spatial road network $\overline{G}$. The output as shown in the dotted line part of Fig. 1(a). Traffic community division can be regarded as the process of road network nodes classification through calculating the classification probability matrix $Y \in R^{N*K*L/2}$ based on predicted traffic information. Finally, we combine the above two processes to predict the structure of time series traffic community.

Table 1. Mathematical notation

Notation	Description
N, F	Number of nodes, feature dimension of nodes
K	Number of divided communities
$L, L/2$	Length of historical and predicted time series
A, D	Adjacency matrix and degree matrix
$X(X_h, X_d, X_w)$	Input data of road traffic network
$Y(Y_h, Y_d, Y_w)$	Classification probability matrix of prediction

Table 1 shows the mathematical notation of this paper.

4 Hierarchical Graph Attention Network

4.1 Model Architecture

We propose a hierarchical graph attention network based on multi-component fusion for predicting the dynamic traffic community, and its architecture is shown in Fig. 2. In this model, we adopt three components to capture the spatial-temporal features of recent, daily and weekly period, and each component constructs a layer graph attention structure through fusing the ASTGCN [4] and the ADGCN, and the two model use independent Loss function for model training. Finally, the output features of each layer are combined to get the prediction output of the model.

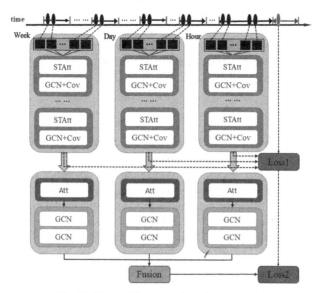

Fig. 2. The architecture of the fusion model

The three components jointly construct hierarchical graph attention network to capture the spatial-temporal features of road traffic network from recent period, daily period and weekly period. And the input traffic data of the three components are the recent segment $X_h \in R^{N*F*L}$, the daily-periodic segment $X_d \in R^{N*F*L}$ and the weekly-periodic segment $X_w \in R^{N*F*L}$ respectively. As shown in Fig. 3, the recent segment X_h is the time series segment directly adjacent to the forecast segment; the daily-periodic segment X_d on the past few days at the same time period as the forecast period; the weekly-periodic segment X_w is the time series segment which have the same week attributes and time intervals as the forecast period. And the time segment $Y \in R^{N*F*L/2}$ represents the predicted segment.

Each component adopts the same network structure to construct a layer graph attention network, which is composed of the ASTGCN and the ADGCN and has an independent classification probability matrix as the prediction output. Finally, the final prediction results are obtained by the three components fusion.

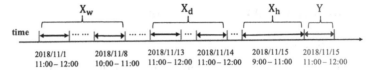

Fig. 3. Three period input time segments and prediction time segment

4.2 ASTGCN

The ASTGCN is the first part of the hierarchical graph attention network to predict traffic information of urban road network. The model is composed of spatial-temporal attention module (STAtt in Fig. 2) and spatial-temporal convolution module (GCN + Cov in Fig. 2), which adaptively capture the spatial-temporal dependence and extract the spatial-temporal features of the adjacent nodes respectively. The two modules jointly constitute a spatial-temporal block, and obtain a larger range of road network spatial-temporal correlation through multi-layer block convolution process to predict traffic information more accurately.

4.3 ADGCN

The ADGCN is the second part of the hierarchical graph attention network for traffic community division. The model uses graph convolution network to extract the features of adjacent nodes related to topological connections, and classifies the adjacent nodes with similar features to divide traffic community based on the traffic information of road network. In addition, the attention mechanism is added to the model to adaptively obtain the correlation of the adjacent nodes to highlight the correlation degree of different nodes and effectively improving the node classification effect (Fig. 4).

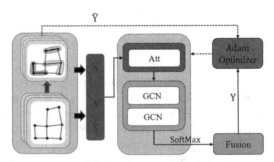

Fig. 4. The architecture of ADGCN. In which, A is the adjacency matrix of the road network, X is the model input data. In addition, we use the community division algorithm to obtain the number of traffic communities K and node classification information $\hat{Y}$.

To obtain the correlation degree of the adjacent node, the attention mechanism proposed [12] is introduced into the ADGCN, which can be written as formula (1):

$$e_{ij} = F\left(W^{(0)}x_i, W^{(0)}x_j\right) \tag{1}$$

where e_{ij} represents the importance of node i to node j, x_i, x_j is the node feature vector, $W^{(0)}$ is the weight matrix, and the function F is the transformation function from the weight matrix to the node importance. To avoid the interference of too much information, the model only pays attention to the neighboring nodes, which is expressed as $j \in N_{ei}$. In Addition, we adopt function softmax to regularize the neighbor nodes to get the attention coefficient between different nodes. Taking the attention coefficient a_{ij} of node i to node j as an example, it can be expressed by formula (2):

$$a_{ij} = softmax(e_{ij}) = \frac{exp(e_{ij})}{\sum_{k \in N_{ei}} exp(e_{ik})} \tag{2}$$

After completing the calculation of the attention coefficient of neighbor nodes, using the nonlinear activation function φ, the output of each node is obtained as:

$$x_i = \varphi\left(\sum_{k \in N_{ei}} a_{ik} W^{(0)} x_k\right) \tag{3}$$

To obtain the classification information of road network nodes, we use double-layer graph convolution network to calculate each nodes partition probability matrix. In the first graph convolution layer, we use the feature matrix $\hat{X}$ of the fusion attention as the input, and renormalize the adjacency matrix A to get the renormalized adjacency matrix $\hat{A}$. $\hat{A}\hat{X}$ can be regarded as the convolving the graphics signal in the spectral domain, the convolution process of the first layer can be expressed as $\hat{A}\hat{X}W^{(1)}$, where $W^{(1)}$ is the learning parameter. To allow the model to obtain the nonlinear relationship, we use the activation function $ReLU$ in the first convolution layer. In the second graph convolution layer, we adopt the same convolution process as the first convolution layer, and take the output of the first convolution layer as the input of the layer, which $W^{(2)}$ is the learning parameter of the second convolution layer. In order to get the each node classification probability matrix, we use the activation function softmax of the second convolution layer. Taking the component of recent period as an example, the output can be expressed as formula (4):

$$Y_h = softmax\left(\hat{A}ReLU\left(\hat{A}\hat{X}_h W^{(1)}\right)W^{(2)}\right) \tag{4}$$

4.4 Multi-component Fusion

We use the method of multi-component fusion to integrate the prediction results of the recent, daily-periodic and weekly-periodic components. In the process of the fusion, we set the parameter matrix to capture the influence weights of different components on the division results and adopt the softmax function to normalize the final predicted classification probability matrix. Finally, by classifying and geographical mapping road network nodes of the same category, we can get the topological structure of future time series traffic community. The above process is shown in Fig. 5.

The multi-component fusion process can be expressed by the formula (5):

$$Y = W_h \odot Y_h + W_d \odot Y_d + W_w \odot Y_w \tag{5}$$

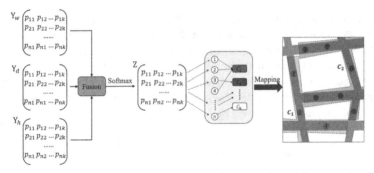

Fig. 5. The left part is the process of multi-component fusion, and the right part is the process of nodes classification and geographically mapping. In which, $1, 2 \ldots n$ are n road network nodes, $C_1, C_2 \ldots C_k$ are k divided traffic communities.

where Y_h, Y_d, Y_w are the classification probability matrix predicted of the recent, daily-periodic and weekly-periodic components respectively. W_h, W_d, W_w are the learning parameter matrix of different components. $\odot$ is the Hadamard product of the prediction probability matrix and the learning parameter matrix.

5   Experiments

5.1   Datasets and Environment

In order to verify the effectiveness of our model, we conducted several comparative experiments on the Xi'an urban traffic dataset of Didi Chuxing. The dataset records the average speed, travel time index (TTI) of 789 sensors in October and November 2018 within the second ring of Xi'an, as well as the topological connectivity of each sensor. The records of each sensor are aggregated into the window of 10 min, and detects 6 times per hour. In addition, we set the dataset of October and November to dataset1 and dataset2, and each dataset is divided into three parts, of which the data of the first two weeks are used as training, the data of the third week are used for verification, and the data of the last week as testing. We use Intel Xeon(R) Silver 4210 CPU @2.20 GHz 64.0 GB GPU GeForce RTX 2080Ti as our experimental configuration and TensorFlow framework as the basic experimental environment to build our model.

5.2   Experimental Details

According to the original data, we carried out data filtering and data completion to obtain the traffic data of the second ring road in Xi'an.

In the stage of model training, we used the Louvain algorithm [13] to divide the traffic community of road network to obtain the classification label of each nodes in different time series, and adapted the number K of current time traffic communities as the number of dynamic traffic communities per time series. In addition, we adjusted the time span of the data by controlling the length of the temporal convolutions. We set the size of the three segments window: $L_h = L_d = L_w = 12$ and the size of predicting window $L_p = 6$ to predict the traffic information over one hour in the future.

For the setting of the model training parameters, the training batch size is set to 16, the epoch of the training phase is set to 200, the learning rate is set to 0.001, and adopt the Adam optimizer for parameter optimization. We take the mean square error (MSE) of the estimator and the real value as the loss function, and minimize it by back propagation to obtain the learning parameters of the model.

5.3 Experimental Results

In order to prove the effectiveness of the prediction model proposed in this paper, we conducted two experiments to verify traffic community division effect of the ADGCN and traffic community prediction effect of the prediction model.

In Experiment 1, in order to demonstrate the effectiveness of the ADGCN, we used accuracy (AC) and normalized mutual information (NMI) as the evaluation indicators of traffic community division effect, and compared it with other methods, including: DCSBM [14], NetMRF [15], PCLDC [16], NEMBP [17], DIGCN [18]. DCSBM and NetMRF methods only use the network topology to divide the network. PCLDC, NEMBP methods also consider the node topology and node attribute information, DIGCN is a community division method based on graph convolution network. The above five algorithms and ADGCN all use the same datasets (dataset1 and dataset2). The comparison results are shown in Table 2.

Table 2. Evaluation results of different methods on the Xi'an traffic dataset

Model	Dataset1		Dataset2	
	AC (%)	NMI (%)	AC (%)	NMI (%)
DCSBM	38.2	10.2	37.1	9.6
NetMRF	39.2	16.5	32.8	17.2
PCLDC	57.4	21.2	50.2	22.6
NEMBP	56.2	18.8	57.9	22.1
DIGCN	58.3	25.1	57.4	25.3
ADGCN	**59.0**	**25.6**	**58.1**	**26.1**

From the evaluation results of each method, it can be seen that the AC of the ADGCN in the two datasets is 59.0% and 58.1%, and the NMI is 25.6% and 26.1%, which are higher than other methods. Among these methods, DCSBM and NetMRF have lower AC and NMI in the two datasets, because the two methods ignore the node features and only focus on the topological connection relationship of nodes. The AC and NMI of PCLDC and NEMBP are lower than the ADGCN's due to the spatial correlation of road network nodes cannot be extracted effectively. In the comparison with the DIGCN, the attention mechanism is added to the ADGCN to effectively extract the correlation between adjacent nodes, so the AC and NMI of our model are higher than the DIGCN. In summary, the ADGCN proposed in the paper has a better effect in traffic community division.

In Experiment 2, we compared dynamic traffic community prediction effect of the prediction model proposed in the paper with attention mechanism and ADGCN + GCN model without attention mechanism to prove the effectiveness of our model, and use modularity Q [19] to measure the traffic community prediction effect. Modularity is an evaluation criterion to measure the community division of complex networks. The greater the modularity, the higher the similarity of nodes within the divided region and the better the regional division effect. The calculation is shown in formula (6):

$$Q = \frac{1}{2m} \sum_{vw} \left[A_{vw} - \frac{k_v k_w}{2m} \right] \delta(C_v, C_w) \qquad (6)$$

The two prediction models adopt the same datasets (dataset1 and dataset2), and predict the dynamic traffic community of four time periods in the two datasets, the four time periods are 6:00–7:00, 9:00–10:00, 14:00–15:00, and 17:00–18:00. The four prediction comparisons are shown in Fig. 6.

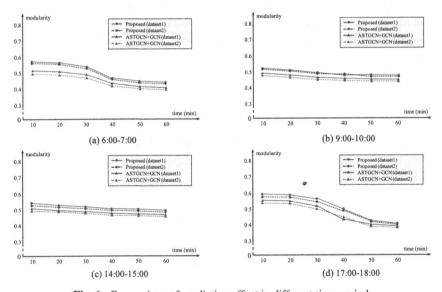

Fig. 6. Comparison of prediction effect in different time periods

From 9:00 to 10:00 and from 14:00 to 15:00, the traffic state of urban road network is relatively stable. The prediction model proposed in this paper and ASTGCN + GCN prediction model have relatively good and stable prediction results. From 6:00 to 7:00 and from 17:00 to 18:00, the urban traffic is in the morning peak and evening peak, the traffic condition of road network changes greatly. The two models have good prediction effect in the short term (10 min–30 min), but in the long term (40 min–60 min), the prediction effect shows a significant decline. In terms of the overall prediction effect, the prediction model proposed in this paper can adaptively obtain the correlation degree of adjacent nodes by fusing the attention mechanism, and effectively distinguish the divided community of adjacent nodes by extracting the correlation information. The model prediction effect is significantly better than that of the ASTGCN + GCN model without attention mechanism, which proves the effectiveness of the model.

6 Conclusion

In this paper, we construct the ADGCN to solve the problem of traffic community division by adding attention mechanism to the double-layer graph convolution network. The model extracts the spatial and temporal features of the road network nodes from the average speed, travel time index to divide the road network into different regions. In addition, we use multiple components to compose a hierarchical graph attention network and construct each component by fusing the ASTGCN and the ADGCN. Finally, combined with the output matrix of each component to represent the dynamic traffic community of each road network node. We conducted multiple comparative experiments on the real-world traffic datasets to prove the effectiveness of the proposed model.

Acknowledgment. This work is supported by National Key R&D Program of China (No. 2017YFC0803300), the Beijing Natural Science Foundation (No. 4192004), the National Natural Science of Foundation of China (No. 61703013, 91646201, 62072016).

References

1. De Montis, A., Barthélemy, M., Chessa, A., Vespignani, A.: The structure of interurban traffic: a weighted network analysis. Environ. Plan. B: Plan. Des. **34**(5), 905–924 (2016)
2. Jiang, B.: A topological pattern of urban street networks: universality and peculiarity. Physica A **384**(2), 647–655 (2007)
3. Wright, C., Roberg, P.: The conceptual structure of traffic jams. Transp. Policy **5**(1), 23–35 (1998)
4. Guo, S., Lin, Y., Feng, N.: Attention based spatial-temporal graph convolutional networks for traffic flow forecasting. Proc. AAAI Conf. Artif. Intell. **33**, 922–929 (2019)
5. Ramezani, M., Haddad, J., Geroliminis, N.: Integrating the dynamics of heterogeneity in aggregated network modeling and control. Transportation Research Board Meeting (2014)
6. An, K., Chiu, Y.-C., Xianbiao, H., Chen, X.: A network partitioning algorithmic approach for macroscopic fundamental diagram-based hierarchical traffic network management. IEEE Trans. Intell. Transp. Syst. **19**(4), 1130–1139 (2018)
7. Kipf, T.N., Welling, M.: Semi-supervised classification with graph convolutional networks (2016)
8. Jin, D., Liu, Z., Li, W.: Graph convolutional networks meet Markov random fields: semi-supervised community detection in attribute networks. Proc. AAAI Conf. Artif. Intell. **33**, 152–159 (2019)
9. Williams, B.M., Hoel, L.A.: Modeling and forecasting vehicular traffic flow as a seasonal ARIMA process: theoretical basis and empirical results. J. Transp. Eng. **129**(6), 664–672 (2003)

10. Jeong, Y.S., Byon, Y.J., Castro-Neto, M.M.: Supervised weighting-online learning algorithm for short-term traffic flow prediction. IEEE Trans. Intell. Transp. Syst. **14**(4), 1700–1707 (2013)

11. Zhang, J., Zheng, Y., Qi, D., Li, R., Yi, X., Li, T.: Predicting citywide crowd flows using deep spatio-temporal residual networks. Artif. Intell. **259**, 147–166 (2018)

12. Zhou, G., Chen, F.: Urban congestion areas prediction by combining knowledge graph and deep spatio-temporal convolutional neural network. International Conference on Electrome-chanical Control Technology and Transportation (2019)

13. Veličković, P., Cucurull, G., Casanova, A.: Graph attention networks. arXiv preprint arXiv: 1710.10903 (2017)

14. Karrer, B., Newman, M.E.J.: Stochastic blockmodels and community structure in networks. Phys. Rev. E **83**(1 Pt 2), 016107 (2011)

15. He, D., You, X., Feng, Z.: A network-specific Markov random field approach to community detection. Proceedings of the AAAI Conference on Artificial Intelligence (2018)

16. Wang, X., Jin, D., Cao, X.: Semantic community identification in large attribute networks. Proceedings of the AAAI Conference on Artificial Intelligence (2016)

17. He, D., Feng, Z., Jin, D.: Joint identification of network communities and semantics via integrative modeling of network topologies and node contents. Proceedings of the AAAI Conference on Artificial Intelligence (2017)

18. Li, Q., Han, Z., Wu, X. M.: Deeper insights into graph convolutional networks for semi-supervised learning. Proceedings of the AAAI Conference on Artificial Intelligence (2018)

19. Blondel, V.D., Guillaume, J.-L., Lambiotte, R., Lefebvre, E.: Fast unfolding of communities in large networks. J. Stat. Mech.: Theory Exp. **2008**(10), P10008 (2008)

ENSTDM: An ENtity-Based Spatio-Temporal Data Model and Case Study in Real Estate Management

Wei Xiong[1][✉], Hao Chen[1], Ning Guo[1,2], Qi Gong[1], and Wenze Luo[1]

[1] National University of Defense Technology, Hunan 410073, Changsha, People's Republic of China
{xiongwei,hchen,guoning10,gongqi12,luowenze12}@nudt.edu.cn
[2] Academy of Military Sciences, Beijing 100091, People's Republic of China

Abstract. The geographic data models are mainly based on geographical entities nowadays. There still exist some problems about these models: separation of the concept and storage for a complete geographic entity (river, railway, etc.), delayed updates in some real-time applications, lack of relationship between geographical entities, difficult management and linkage update for multi-source heterogeneous data, and so on. A unified data model for geographic entities and geographic events is designed to respond to the problems. In this model, data can be organized based on the basic granularity of the conceptually complete geographic entity. Geographical events is fully described using geometry, attribute and relationship information, and interaction between geographical entities and geographical events is achieved. Based on GeoJSON, GeoEntityJSON and GeoEventJSON are designed to implement the physical model of the data. Taking a residential area in a certain city in China as an example, a real estate management model is established. The data model is used to realize the storage management and visualization of real estate.

Keywords: Data model · Geographic entity · Geographic event · Real estate management

1 Introduction

There always exists a gap between the conceptual understanding and the practical application of geographic data. For example, rivers in the existing GIS softwares are stored as multiple river sections instead of the river as a whole; the same river has different data sources under different application backgrounds, which results in data redundancy. When designing our new generation of geographic infrastructure, a solution to the problems described above is to store

---

Supported in part by the National Natural Science Foundation of China under Grant No. 41471321. and No. U19A2058.

G. Pan et al. (Eds.): SpatialDI 2021, LNCS 12753, pp. 27–42, 2021.
https://doi.org/10.1007/978-3-030-85462-1_3

the geographic world directly using the conceptually complete geographic entities. Although in recent years the idea of geographic entity oriented modeling has been widely implemented [2,11,14], many issues have not yet come to a consensus.

Many researchers tried to find a general spatio-temporal data model: Event-based Spatiotemporal Data Model (ESTDM) [3] by Peuquet used a temporal-based method by recording the 'event list' on a time-line. Dynamic geo-networks [5] introduced the relations between events and states to record the dynamic characteristics of geographic phenomenon. However,the model did not extend to deal with multiple attributes changes. FEM [12] was based on "state-event-state" approach and applied the object-oriented model to represent both geographic entities and their changes. An obvious performance problem was that all the changes were recorded in versions, thus frequently updated sensor data will then generated redundant versions. [7] designed a real-time spatio-temporal data model for dynamic geographic objects and dynamic process simulation, but it was mainly for fast-changing objects without considering the possible relations among them. Geographic information extraction and geographic knowledge graph has been researched in [13], which indicated that semantic support is needed in data model nowadays. [1] devised a conceptual data model for dynamic changes expression of spatio-temporal object and its association, which incorporated not only dynamic changes of spatial characteristics and attribute characteristics of geographic entities, but also the various associations of geographic entities over space and attribute. The data model, unfortunately, lacked any possible case studies.

There still exist the following four problems about these models: Firstly, the concept of a complete geographic entity (river, railway, etc.) is divided into multiple geographical objects for storage, which violates the concept integrity and makes it difficult to extract and update relevant information of the entity. Secondly, the real-time changes in the attribute information of geographic entities cannot be reflected in the data. Thirdly, the geographic data model only contains spatial information and attribute information of geographic entities, and does not reflect the relation information between geographical entities. Fourthly, multi-source heterogeneous data cannot achieve unified management and linkage update. At the same time, based on traditional geographic entity data model, the current map can only support the search for geographical entities. As the user's demand for event information increases, a geographic event library parallel to the geographic entity library should be established.

These issues are addressed in this paper with an entity-based spatio-temporal data model (ENSTDM), which is designed to organize the multi-source, spatial-temporally changed data. What we do in this paper, in summary, is to design a data model for geographic database system aiming to efficiently record the history and predict the future of geographic entities. In this paper we devise a general spatio-temporal data model based on geographic entities and geographic events, implement its data structure using GeoJSON [6,10] and apply it the real estate management.

2 Data Model

In the existing GIS software, line data such as rivers, roads, and railways are usually stored in segments in the database. Figure 1 shows how the data of Xiangjiang River is organized in the spatial database, numbers represent feature id of each segment. In this way, a geographic entity with a complete concept is cut into multiple spatial features, and it is difficult to extract the complete information of the entire entity. At the same time, it is difficult to establish an association between relevant thematic information and geographic entities. This type of segmented simple geographic entity needs to integrate the geometric information, attribute information, and related information of the separated features. The integration methods are different: For example, geometric information should integrate multiple line segments into a complete line segment, attribute information such as length should be added numerically, and related information such as tributary information should be merged.

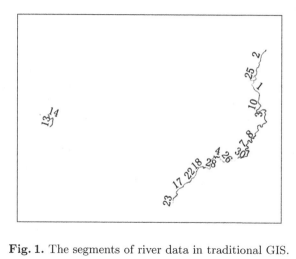

Fig. 1. The segments of river data in traditional GIS.

The geographical entity refers to the natural or artificial objects which are relatively fixed in space and have the common nature (administrative divisions, rivers, communities belong to geographical entities, while wind, cars, and people do not). Geographic events refer to events related to geographical entities occurring at a certain time and a certain location. The spatio-temporal data model based on geographical entities and geographical events consist of two parts: geographical entities and entity relations, geographical events and events relations.

Geographical entities are closely related to geographical events as shown in Fig. 2. A geographical event is involved by one or more geographical entities, which may be automatically generated by a change in a geographical entity and it may also be a bulletin issued by a professional organization. Geographic entities can be generated, deleted, or updated by *registers* of geographic events.

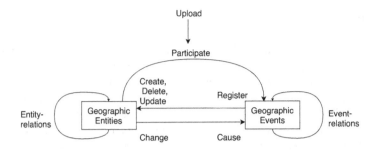

Fig. 2. The interaction between geographic entity and geographic event.

2.1   Geographical Entity and Entity Relations

In this paper, the data model of geographic entities takes the concept of complete geographic entities as the basic granularity, including its spatial information, attribute information and relation information as shown in Fig. 3 and Fig. 4. The formal description of geographic entities is defined in [8]. As for spatial information, it should be taken into account that the same geographic entity presents different geometric forms in different application scenarios. An adaptive hilbert Geohash (AHG) meshing and coding method [9] is adopted to identify and index geographic entities. AHG can adjust the hierarchical grid level according to the geometry attributes of the geographic entities. Also, the geometry of the same geographic entity may change at different time, thus spatial information should be stamped with type and time.

For more accurately representing entities in real world, geographic entities are divided into simple geographic entities (SGE) and complex geographic entities (CGE) based on geometric shapes. SGE is described by a point, line or polygon, while CGE is a new geographic entity concept proposed here. Geometrically, a CGE is a composition of points, lines and polygon. In essence, it can be a composition of multiple homogeneous SGEs and CGEs, or that of multiple heterogeneous SGEs and CGEs, to form a geographic entity with a complete conceptual structure. For example, a SGE with the polygon shape of the Yangtze River, multiple SGEs of docks, navigation aids, etc. form a CGE of the Yangtze River, as shown in the Table 1. SGEs of the Changsha, Wuhan, Hefei, Nanjing and Xi'an campuses of National University of Defense Technology together form the CGE of the National University of Defense Technology. The reason for subdividing CGEs into homogeneous CGEs and heterogeneous CGEs is to determine whether to retain or merge the attributes of the entities.

The geographic world, if we think of it in an abstract way, basically consists of two parts: the geographic objects existing in the real world and their relations. From a perspective of set theory, it can be expressed as Eq. 1 where E, O, R denote the geographic entities set, the geographic objects set and the geographic relations set respectively.

$$E = \{\langle o, r \rangle \mid o \in O, r \in R\} \tag{1}$$

Table 1. Geographic entities classification

Classification		Description	Encoding	Example
SGE		Traditional features described by a point, line or polygon	0	Point entity: city Line entity: river Polygon entity: lake
CGE	Homogeneous	Composition of multiple similar SGEs or CGEs	1	A university composed of different campuses
	Heterogeneous	Composition of multiple SGEs or CGEs which have different conceptual structures	2	A river is composed of different types of piers, navigation aids, tributaries and other geographical entities

Before we look into the geographic entities set, we need to think about the fact that both geographic objects and geographic relations keep changing. A record of these changes, for the new generation of GIS, is a must when describing geographic entities, and this relies greatly on well-defined time. OWL-Time [4], a part of the W3C Semantic Web Activity, gives a good example of expressing rich temporal information using relations and classes. One of the most important pairs of concepts are:

- Instant: a precise moment of time. According to [16], time $\mathbb{T}$ is a universe, which can be mapped to a one-dimensional real Euclidean space denoted by $\mathbb{R}$. Assume that $(\mathbb{T}, <)$ is a strict total order, then Instant is defined as follows:

$$instant = \{t | t \in \mathbb{R} \cup \{\emptyset\}\} \tag{2}$$

- Interval: a sustained period of time. Mathematically, it can be defined in this way [16]:

$$interval = \{(s, e, f_c, f_e) | s, e \in instant, f_c, f_e \in bool, s \leqslant e\} \tag{3}$$

where s, e are the start and end time of a certain interval, f_c, f_e are the flags indicating whether the end is bounded. For example, $(t_1, t_2, 0, 1)$ represents the time interval $(t_1, t_2]$. Another pair of concepts, not mentioned in OWL-Time, are frequently discussed in temporal GIS [15]:

- Transaction time: also called database time or system time. This is the time when data is inserted into the database, usually in a form of instance.
- Valid time: also called event time or real-world time. This is the time when the data is valid. It could be either an instance or an interval .

In traditional geographic data models, attribute information is fixed. In order to better record the change of geographical entity attributes, we can divide the attributes into three parts according to the frequency of change: stable attributes, steadily changed attributes and rapidly changed attributes. The rapidly changed

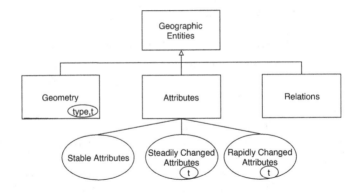

Fig. 3. Geographic entities.

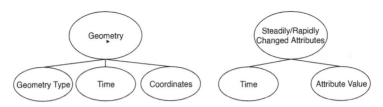

Fig. 4. ER diagram corresponding to Fig. 3.

attributes are mainly for the real-time updates of sensor data. Time should be added to attribute information in order to record the changes.

Geographic entities can be divided into the different categories according to Fig. 3. In this way, we can almost express various real-world geographic entities. For POI data or real estate data, the geometry type and time attribute will both not change during a period of time. For sensor data such as air quality monitor, the geometry type (spatial attribute) will not change, but the time attributes will steadily change. For the check-in data of social networks, the spatial and temporal attributes will change rapidly.

Part of the relational information in traditional geographic data models is missing. The main reason is that "relevance" is an overly broad concept, and there is a certain connection between any two entities in the world. In this paper, we apply object-oriented ideas to define the relations between geographical entities, which is presented in Fig. 5, including "tag", "dependency" "components", "member" and "reference".

- The "tag" relation is designed to specify the category of geographical entities (concept entity), which helps to build a bridge between the domain experts and the data production personnel.
- The "dependency" relation is designed to specify the entity at the higher level. This relation can be used to realize the linkage update of the data (for example, the population of a block depends on the population of the

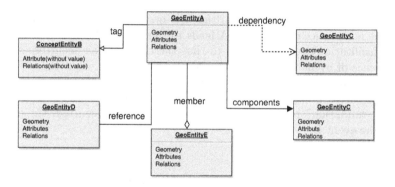

Fig. 5. The relations between geographic entities.

residential district within it, and the two can be linked to achieve synchronous changes).

- The "components" relation is to express the relationship between geographical entities and their components. Once the geographical entity is not existing as a whole, its components are also deleted (For example, a building as an entity is composed of multiple "houses". Once the building no longer exists, the house entity in the building no longer exists.).
- The "member" relation is also to represent the relationship between geographical entities and their parts, but unlike the "components" relationship, even as a whole geographic entity does not exist, its members still exist (For example, a block is composed of multiple houses. The block does not exist if merged to other blocks. The houses in the block can still exist.).
- The "reference" relation is mainly related to multi-source heterogeneous data, such as multimedia data, RDF associated data and so on.

2.2 Geographical Events and Their Relations

Unlike geometry and attributes, geographic events and rules are more ambiguous because they incorporate a great deal semantics. In order for a better understanding by the computer and also a match for our perception of reality, here we present the formal and interchangeable description of the geographic events and rules by borrowing the ideas from semantic web.

It is necessary to design a standard data model for geographic events and establish a geographic event library parallel to the geographic entity library.

In this paper, the geographic event data model takes a single geographic event as the basic granularity, which contains geometry information, description information and relation information. The geometry part is the same with that in geographic entities. The description information can be either text information or multimedia data or other types of data. Relation information refers to the relationship between geographic events. This paper defines the relations between events as shown in Fig. 6, which are "initiation", "facilitation", "blocking" and "termination".

- The "initiation" relation refers to that the event B has not occurred before, and the occurrence of the event A leads to the occurrence of the event B.
- The "facilitation" relation refers to that the event B has occurred, and the occurrence of event A promotes the development of the event B
- The "blocking" relation refers to the event B has occurred, and the occurrence of the event A hinders the development of the event B.
- The "termination" relation refers to the event B has occurred, the event A stops the event B.

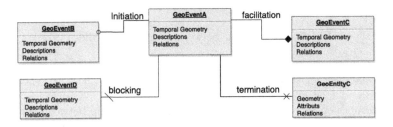

Fig. 6. The relations between geographic events.

These relations are helpful for tracing the semantic relations between events, realizing the linkage update of event information, and providing semantic support for reasoning. What's more, events contain a wealth of information so that answering questions like: Why did that event happen? What is the result of that event? What is the whole process of that event?

2.3   Data Model Implementation

For each geographic entity, it should include the following parts of information: UUID (universal unique identification), as the unique identifier of a geographic entity, is the primary key of each geographic entity; Spatio-temporal reference which explain the reference system adopted by the spatio-temporal data; Geometry indicates the shape and location of geographic entities; Properties record various attributes and attribute values of geographic entities; Description and related information associated with geographic entities.

The key information is UUID, which can be defined as:

$$UUID = Prefixcode + Classificationcode + Sequencecode + Locationcode$$

The prefix code is defined in Table 1. The classification code should be determined by domain experts. The sequence code is generated in random order, and the location code uses AHG encoding method [9]. The following takes a typical linear trajectory object, a single Beijing taxi trajectory as an example to illustrate the AHG encoding process. The selected example trajectory is the driving route from Tiananmen Square in Beijing to the bus station of the Academy of Military Sciences. The latitude and longitude of the space rectangular outer frame MBR is as follows:

$$lon\_min = 116.261669°$$
$$lat\_min = 39.907543°$$
$$lon\_max = 116.397593° \tag{4}$$
$$lat\_max = 40.005935°$$

The spatial range is as 5:

$$\Delta lon = lon\_max - lon\_min = 0.135924°$$
$$\Delta lat = lat\_max - lat\_min = 0.098392° \tag{5}$$

Then the level of AHG is:

$$\frac{360}{2^{l_{adapt}+1}} < max\{\Delta lon, \; 2 \times \Delta lat\} \le \frac{360}{2^{l_{adapt}}}, \; l_{adapt} \in \mathbb{N}^+ \tag{6}$$

The center point coordinates of the trajectory MBR are (116.329631°, 39.956739°). According to the calculation process in [9], use it as the input data of GeoHash encoding, and use the Hilbert space filling curve as the grid traversal sequence. Finally get the binary code of the driving trajectory It is: 11010100110011100011, and its Base32 encoding result is wtdd. Note that the spatial grid size of the current coding level is about 39 km, which matches the scale of the spatial trajectory object (25 km), indicating that AHG can better reflect the location and size characteristics of two-dimensional spatial object.

For example, Beihai Park in Beijing is a simple geographic entity here, with the classification prefix code 00. The classification code is 120525. There are no other parks in the space of this code, so its sequence code takes the default value of 0. The location code length of the point entity is set to 8 directly. The longitude and latitude of Beihai Park is (39.928167°, 116.389550°), and the location code is wx4g0s8q. Finally, the geographic entity code of Beihai Park is 00120525000wx4g0s8q.

Logical model is software-dependent and should be expressed in terms of the language of a specific DBMS. Relational data modeling and object-oriented data modeling are the most common techniques, while in this paper the latter is applied due to its great power of expressiveness and efficiency. For an object-oriented logical model design, classification should be the first step. We can refer to some standard documents released officially, on which the classification of geographic entities is given. Then the attributes and operations can be defined in the class and contained within all the instances of that class. The most important concepts in object-oriented technology are inheritance, encapsulations and polymorphism, which can be embodied in the logical model in the following way:

- Inheritance: the attributes and operations of a geographic entity can be inherited from its superclass. This reduces data redundancy and facilitates synchronous updates.

- Encapsulations: it allows the geographic entities reusable by shielding their internal data and operations. This also guarantees the security of some sensitive data.
- Polymorphism: the same geographic entities can have different geographic representation. Polymorphism offers a solution to accurately represent for any granularity and application of interest.

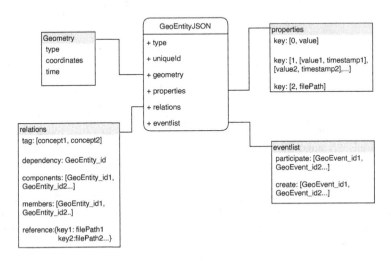

Fig. 7. GeoEntityJSON data structure.

GeoJSON is a commonly used lightweight geospatial data exchange format [6]. The "geometry" field and the "properties" field store the spatial information and attribute information, respectively. In this paper, the implementation of the data model is based on the GeoJSON format with the expansion of geographical entities and geographical events, named GeoEntityJSON and GeoEventJSON. They are designed to formulate semantic specifications to achieve unified management and exchange of data. The data structure is shown in Fig. 7 and Fig. 8. Both of them are designed according to the previous analysis. It should be noted that 1) *type* should be the same with that in GeoJSON; 2) *uniqueid* is a code that uniquely determines geographic entities or geographic events; 3) the *properties* in GeoEntityJSON and *descriptions* in GeoEventJSON has three types: the first is invariant property (flag = 0), the second is steadily changed property (flag = 1), the third is rapidly changed property (flag = 2), which is mainly for the data stored in other locations; 4) both GeoEntityJSON and GeoEventJSON should record the related events and entities respectively, with the aim of recording the interaction of geographical entities and geographical events.

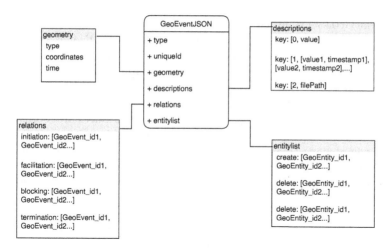

Fig. 8. GeoEventJSON data structure.

A typical geographic entity represented in geojson format is as follows:

```
1  {
2      "type": "Feature",
3      "UUID": "001602010h0",
4      "geometry": {
5          "type": "MultiLineString",
6          "coordinates": ["..."]
7      },
8      "properties": {
9          "NAME": "Yangzi river",
10         "GBCODE": 21011,
11         "LENGTH": 64.869,
12         "LEVEL": 1,
13         "transact_time": "18/03/2018",
14         "valid_time": "18/03/2018"
15     },
16     "relation": {
17         "flag": 5,
18         "name": "Tributary",
19         "entities": ["..."]
20     },
21     "meta": {
22         "note": "null",
23         "precision": "null",
24         "produce_time": "null",
25         "producer": "null",
26         "security_level": "null"
27     }
28 }
```

Once the GeoJSON files are created, document database is needed. MongoDB is a suitable open-source NoSQL database, in which GeoJSON files can be stored with indexes and queried in an efficient way. As for the issue of representing relationships, the database Neo4j is preferred. It is a graph database which functions well when storing relationship. There are several points that should be noted: The geographic entities belonging to the same category are stored together. For geographic objects, geometry and steadily changed attributes are stored in versions and only the changes related to specific geographic objects are recorded. That has the obvious advantage of avoiding storage of redundant information. The description of events and rules can be extracted to do analysis.

The basic idea for geographic entity storage is "conceptually centralized while physically distributed", which is shown in Fig. 9.

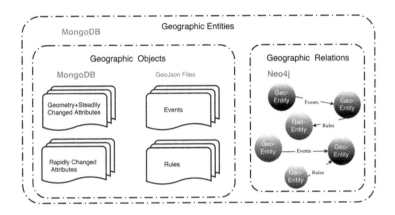

Fig. 9. Physical storage model.

3   Real Estate Use Case

Now housing data on the map is mainly based on buildings or residential areas (interest points), but the real estate department own details about real estate meticulous to households. In order to display these rich information more intuitively on the map, a three-dimensional entity data model of real estate management should be set up to realize the storage, management and visualization of the housing information. We apply our data model described in to achieve this. The entity data model 1) includes the real estate entities and the real estate events; 2) involve hierarchical display from residential areas to buildings to households; 3) adapt the model to three dimensional display.

For example, a community entity should also include buildings, and the buildings should include houses of different types and storey heights. The existing data of geographic objects cannot reflect this hierarchical relationship. One of the advantages of our data model is that it can describe this complex geographic entity. The original experimental data comes from the CAD design drawing of a community in Xupu County, Hunan Province. The traditional method is generating these data into multiple layers for management. We transform the data and obtain the complex entities of the community.

Real estate entities contain three hierarchical levels of entities, namely, residential areas, buildings, households. Each entity contains spatial information, attribute information and relation information. We display these there levels respectively in Fig. 10, 11, 12.

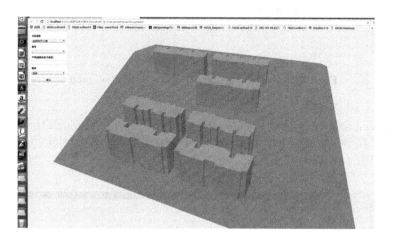

Fig. 10. Residential area display on map.

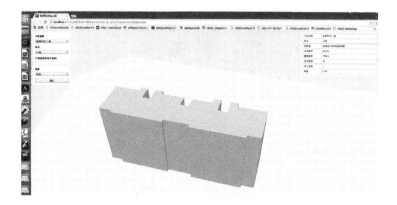

Fig. 11. Building within residential area display on map.

Real estate event mostly include housing sales, maintenance and demolition and so on. Each event contains spatial information, description information and relation information. We set the construction, sales and maintenance of

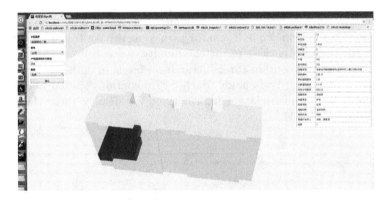

Fig. 12. Household within building display on map.

one building as an example to show the event in Fig. 13. Through the detailed records of these information, a full cycle management of residential housing can be achieved.

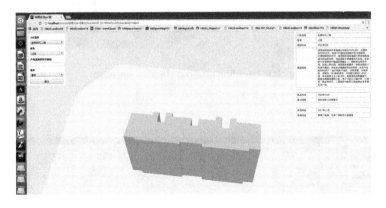

Fig. 13. Real estate event display on map.

4   Conclusions

In this paper, we design the ENtity-based Spatio-Temporal Data Model (ENSTDM), aiming to represent the geographic world in a dynamic and intellectual way. The contribution of the model includes the following points: 1) Data is organized based on the basic granularity of the conceptually complete geographic entity. 2) The expansion of time is demonstrated in the attribute. 3) Relations between geographical entities is established in an object-oriented manner. 4) The goal is changed from the management of the data itself of multi-source heterogeneous data to the management of the data source. 5) Geographical events is fully described using geometry information, description information and relation information, and interaction between geographical entities and geographical events is achieved. The second contribution is that based on GeoJSON, a commonly used lightweight geographic data exchange format, GeoEntityJSON and GeoEventJSON are designed to implement the physical model of the data. On top of this data model, we devise GeoEntityJSON and GeoEventJSON to implement the physical management of geographic entities and geographic events. Finally we apply the model to real estate entities and real estate events, achieving the storage management and visualization of these information. In the future, the dynamic sensor data of the real estate can be used to monitor the housing situation in real time and eliminate the hidden dangers.

Compared to existing works, data from multiple sources are organized as object-oriented, with geographic entity being the basic element. By integrating the attributes changes within the entities, their history is stored and queried in an more efficient way than the traditional entity model. Moreover, spatial-temporal changes are further classified, making expression of geographic entity more accurate and understandable. In addition, the formal description of geographic rules and events are incorporated in the data model, so that new geographic objects or new relationships can be predicted based on that. The work is meaningful because it provides a solution to the new generation of GIS. We expect that this will further foster the construction of geographic infrastructure.

However, there still exists potential improvements including: Moving objects should be considered in the data model. For the semantic part of the data model, ontology should be introduced. Efficiency should be pay more attention about physical model. More query and analysis mode should be explored.

References

1. Cheng, B., Guan, X., Xiang, L., et al.: A conceptual data model for dynamic changes expression of spatio-temporal object and its association. J. Geo Inf. Sci. **19**(11), 1415–1421 (2017). (in Chinese)
2. Devillers, R., Desjardin, R., De Runz, C.: Imperfection of Geographic Information: Concepts and Terminologies, Chap. 2, pp. 11–24. Wiley (2019). https://doi.org/10.1002/9781119507284.ch2
3. Peuquet, D.J., Duan, N.: An event-based spatiotemporal data model (ESTDM) for temporal analysis of geographical data. Int. J. Geogr. Inf. Syst. **9**(1), 7–24 (1995). https://doi.org/10.1080/02693799508902022
4. Espinozaarias, P., Povedavillalon, M., Garciacastro, R., Corcho, O.: Ontological representation of smart city data: from devices to cities. Appl. Sci. **9**(1), 32 (2018)
5. Galton, A., Worboys, M.: Processes and events in dynamic geo-networks. In: Rodríguez, M.A., Cruz, I., Levashkin, S., Egenhofer, M.J. (eds.) GeoS 2005. LNCS, vol. 3799, pp. 45–59. Springer, Heidelberg (2005). https://doi.org/10.1007/11586180_4
6. Gillies, S., Butler, H., Daly, M., Doyle, A., Schaub, T.: The GeoJSON format. RFC **7946**, 1–28 (2016)
7. Gong, J., Li, X., Wu, H.: Spatio-temporal data model for dynamic GIS. J. Survey. Map. **3**, 226–232 (2014). (in Chinese)

8. Gong, Q., Guo, N., Xiong, W., Chen, L., Jing, N.: A spatio-temporal data model of geographic entities. In: 2018 26th International Conference on Geoinformatics, pp. 1–6 (2018)

9. Guo, N., Xiong, W., Wu, Y., Chen, L., Jing, N.: A geographic meshing and coding method based on adaptive Hilbert-geohash. IEEE Access **7**, 39815–39825 (2019)

10. Horbiński, T., Lorek, D.: The use of leaflet and GeoJSON files for creating the interactive web map of the preindustrial state of the natural environment. J. Spat. Sci. **31**, 1–17 (2020)

11. Jiang, J., Huang, W., Lu, W., Zheng, X.: Research on entity-based data modeling for national geo-spatial information service platform. Geogr. Inf. World **7**(4), 11–18 (2009). (in Chinese)

12. Lohfink, A., Carnduff, T., Thomas, N., Ware, M.: An object-oriented approach to the representation of spatiotemporal geographic features. In: Proceedings of the 15th Annual ACM International Symposium on Advances in Geographic Information Systems, p. 35 (2007)

13. Lu, F., Yu, L., Qiu, P.: Geographic knowledge graph. J. Geo Inf. Sci. **19**(6), 723–734 (2017). (in Chinese)

14. Sun, B., Shufean, A., Sun, F.: A geographic information system framework for data visualisation in wireless sensor networks. Int. J. Sensor Netw. **19**(1), 51–61 (2015)

15. Worboys, M.: Object-oriented approaches to geo-referenced information. Int. J. Geogr. Inf. Syst. **8**(4), 385–399 (1994)

16. Yuan, Y., Gao, Y.: Object-oriented spatial temporal data model and its implementation technology. Geogr. Geogr. Inf. Sci. **24**(3), 41–44 (2008). (in Chinese)

Intelligent Extraction Method of Inertial Navigation Trajectory Behavior Features Considering Road Environment

Xiang Li[1], Wenbing Liu[2($\boxtimes$)], and Qun Chen[1]

[1] Information Engineering University, Zhengzhou 450001, Henan, China
[2] Army Logistics Academy, Chongqing 401311, China

Abstract. Inertial navigation systems play an important role in areas without satellite signals. However, the inertial navigation error changes over time. Therefore, it is essential to use external data sources to correct the trajectory data, and feature extraction is required for error correction. At present, the inertial navigation trajectory consists primarily of track points and segments, with relatively simple features, no time-series data, and no spatial attributes, such as elevation and speed. In this paper, a convolutional neural network (CNN) is established based on deep learning theory and the rich behavioral features of the inertial navigation trajectory. A feature classification model based on the CNN and gated recurrent unit (GRU) is proposed to extract features from the real-time inertial navigation trajectory. First, preprocessing of the vehicle data obtained by the inertial navigation system is performed to filter redundant and invalid data. Subsequently, the trajectory features are categorized according to the motion trend and time-series information. The trend feature vector CNN and the time-series feature vector are used as inputs to the CNN and GRU, respectively, and the trajectory model with the behavioral features is established. Finally, the long short-term memory (LSTM) model, which is prone to overfitting when many parameters are used, is improved, and the two feature vectors are input into the CNN model and GRU model for data fusion to ex-tract the behavioral features. A typical inertial navigation trajectory dataset of the Miyun area in Beijing in December 2015 is used to extract the behavioral features. The experimental results are compared with traditional feature ex-traction and neural network classification methods. The results show that the proposed method outperforms the other methods, with a feature extraction ac-curacy of 91.44%. The method shows excellent performance for extracting the behavioral features of the inertial navigation trajectory in a region with variable elevation and velocity.

Keywords: Inertial navigation system · Trajectory behavior feature · Feature vector · Deep learning · Blending feature neural network · Feature extraction

1 Introduction

In navigation and positioning applications, satellite signals experience interference by various factors, resulting in unreliable positioning results. An inertial navigation system

© Springer Nature Switzerland AG 2021
G. Pan et al. (Eds.): SpatialDI 2021, LNCS 12753, pp. 43–56, 2021.
https://doi.org/10.1007/978-3-030-85462-1_4

is an autonomous navigation system that does not rely on external information and does not radiate energy to the outside world. However, its positioning error accumulates over time; thus, external auxiliary data are required for error correction. The correction of the inertial navigation positioning error has been the focus of research domestically and internationally because accurate extraction of the trajectory features is necessary to correct the error and improve the accuracy.

Many scholars globally have studied methods for extracting the trajectory features. For example, the Douglas-Peucker (DP) algorithm and its derivative algorithm have been widely used as a classical vector feature extraction algorithm. However, it requires iterative calculations, resulting in low efficiency for complex data [1]. Reference [2] proposed the concept of the trend feature and performed feature ex-traction based on the trend change. In reference [3], geo-hash spatial coding was used as the breakthrough point, and a clustering algorithm was used to extract the feature points where the angle changed. Wu et al. [4] avoided the problem of track aggregation in different driving directions using the track point migration method and accurately extracted geometric features. Zhu et al. [5] constructed a semantic feature library, analyzed the dependency relationship between features, and accurately identified the behavioral features of the trajectory. The above methods provided high extraction efficiency and accuracy but are based on mathematical models. The parameter threshold in the extraction process is estimated or is an empirical value, and new data cannot be integrated into the model, resulting in limitations.

Due to the increased use of artificial intelligence in recent years, many machine learning and deep learning algorithms have been used to extract trajectory features, and data-driven feature extraction methods have become popular. Zheng [6] summarized research methods for trajectory data mining and feature recognition and divided trajectory feature extraction into two categories. Lu et al. [7] used a generative adversarial network for deep learning to extract lane-level features. Yu et al. [8] and Stavros et al. [9] used machine learning algorithms to determine the thresholds of feature change of instantaneous data. Cui et al. [10] extracted deep features from trajectory data using an automatic encoder and combined the data with the original features as the input of long a short-term memory (LSTM) network. Lv et al. [11] used trajectories transformed into two-dimensional images as input into a convolutional neural network (CNN) for feature extraction and trajectory prediction. In addition, a clustered hidden Markov model, the classic LSTM model, multi-layer perceptron, and a bidirectional recurrent neural network were used for trajectory feature extraction, providing excellent results [12–15].

The above studies focused on the improvement of neural networks but did not consider multiple behavioral features of the inertial navigation trajectory, making it difficult to analyze the influence of the external environment on the behavioral features and subsequent motion direction. Therefore, this paper integrates the rich parameters and features of the inertial navigation trajectory and establishes a hybrid self-learning method to extract the behavioral features or the inertial navigation trajectory.

2   Description and Extraction of the Behavioral Features of the Inertial Navigation System Trajectory

2.1   Related Work

Trajectory modeling of vehicle data is used to determine the motion features. Most studies focused on modeling the motion trajectory using satellite navigation (GPS) and can be divided into trajectory point-based and trajectory segment-based research [16].

(1) Methods based on trajectory points. Generally, the motion process of vehicle data is discretized into a time series of trajectory points, where each time point describes the behavior of the vehicle based on location information and the heading angle, speed, and mileage [17]. Generally, the grid method is used to divide the road area into blocks to establish the mapping relationship between the track points of the vehicle and the grid. The unit is a single grid, and the complex track data are divided into segments according to the minimum description length. This method does not consider the continuity of the motion of the vehicle, and the local behavioral features cannot be obtained. All tracks in the grid cell are analyzed, and only general information can be obtained [18]. The inflection point detection algorithm [19] improves the grid. Starting from the angle, the algorithm divides the trajectory data into smaller segments for processing. This method also improves the accuracy of trajectory clustering but does not con-sider the spatial attributes of the trajectory data. The trajectory is divided based on the position and geometric features of the trajectory data [20]. The spatial attributes of the trajectory data are considered, but the feature points are discrete, leading to inadequate results. The track data analysis method based on the starting point to the endpoint [21] considers the relationship between the area and the track data distribution. This method improves the speed of data division, but the points are discrete, resulting in low accuracy for complex road conditions.

(2) Methods based on trajectory segments. Numerous studies used the trajectory segment method to describe the motion of mobile vehicles. The vehicle's track is divided using various methods, such as the analysis of the heading angle, piece-wise linear segmentation [22], and a sliding window [23] to obtain a set of track segments with different lengths. Each segment describes the motion state of the vehicle (velocity, acceleration, attitude, coordinates, etc.) and changes in the start and endpoints. Since the vehicle is driving on a road, and the behavior mode changes depending on the road state, the track segment still has a variety of motion states [24–27], such as straight travel, turning around, and lane changes. Trajectory segments can describe the local motion features of vehicles better than trajectory points; however, different segmentation methods produce different motion features, affecting the subsequent behavior patterns of the mobile vehicle.

Regardless of the method, the geometric structure is typically used to model the movement of the vehicle at different scales and granularity to detect the similarities and differences in the trajectory of the vehicle. However, the two methods lack descriptions of the behavioral features of the vehicle's movement and the semantic aspect of the motion.

In addition, the description of the trajectory is based on a two-dimensional plane, and elevation changes are not considered in the behavioral features.

Therefore, this paper uses the behavioral features to model the inertial navigation system trajectory to describe the vehicle's motion state accurately.

2.2 Extraction of the Inertial Navigation Trajectory's Behavioral Features

When vehicles stop moving, measurement errors and other reasons will produce noise and redundant information in the inertial navigation trajectory data, the accuracy of the subsequent trajectory feature extraction.

The data used in this study are vehicle trajectory data obtained in the Miyun area in Beijing in December 2015. The collection interval is 100 ms, and each record has seven fields, as shown in Eq. (1). p_i represents the position, elevation, heading angle, velocity, acceleration, and time of the inertial navigation output at time i.

$$p_i = (x_i, y_i, z_i, h_i, v_i, a_i, utc_i) \tag{1}$$

The trajectory features of vehicles traveling on the road have a one-to-one correspondence to the road data. Since multiple data sources and high-precision maps are available, the track features can be matched to the road features, and data mining can be used. These features often change during the vehicle's movement. They include shape and topology features (turning, intersection, etc.) based on the coordinates (Fig. 1), elevation features (viaduct, underground tunnel) due to the use of three-dimensional data (Fig. 2), and miscellaneous features (speed limit sign, traffic light, sidewalk) based on the traffic rules and driving habits (Fig. 3).

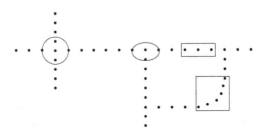

Fig. 1. Geometric features based on shape and topology

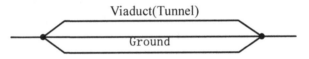

Fig. 2. Elevation features

Fig. 3. Features based on traffic signs

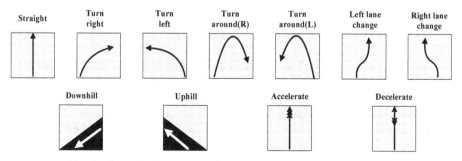

Fig. 4. The 11 behavioral features of the inertial navigation trajectory

The following feature types are selected in this study. Six motion behavior patterns and 11 behavioral features of the vehicle are analyzed, as shown in Fig. 4.

Therefore, the trajectory features consist of multiple factors:

Behavioral feature = geometry feature + elevation feature + velocity feature

The inertial navigation trajectory data contain temporal, spatial, and attribute information, which does not express the behavioral features of the vehicle. Accurate classification of the behavioral features can only be obtained after this information is transformed into data suitable for the classification model. Therefore, an important step is to construct the feature vectors [28] before using the classification model. Existing classification methods analyze the trajectory features using probability statistics of the extracted attributes, such as the mean, variance, and median [29]. However, the disadvantages of the statistics are that manual calculation is needed, and deep features hidden in the data cannot be extracted.

We use the concept of the trend feature [2] and calculate the trend feature vector as the input data of the extraction and classification model. This model analyzes the features of the inertial navigation trajectory and the relationship between parts of the trajectory. This approach is conducive to learning and mining the basic and deep features of the trajectory for accurate extraction and classification of the behavioral features.

2.3 Construction of Trend Feature Vector

Research on traditional classification algorithms has shown that the velocity, acceleration, and heading angle are suitable parameters to distinguish the geometric features of the trajectory. We include elevation as a one-dimensional vertical attribute. In this study, the cumulative changes in the velocity v, acceleration a, heading angle h, and elevation z in a given trajectory interval are used to obtain the trend's eigenvector.

The calculation method of determining the cumulative change in the trend ΔTr between this and the previous interval is as follows:

$$\Delta Tr_i = \begin{bmatrix} \Delta h_i \\ \Delta v_i \\ \Delta a_i \\ \Delta z_i \end{bmatrix} \tag{2}$$

$$\Delta h_i = h_i - h_{i-1} \quad (i \geq 1) \tag{3}$$

where h_i, v_i, a_i, and z_i (I = 1,2, n) represent the heading angle, velocity, acceleration, and elevation, respectively, of the ith track point in the track segment.

The eigenvector matrix of the trend has 50 dimensions; thus, 50 points are used as a sub-interval. The positive and negative values of the cumulative change of the trend represent the trend direction relative to the current state of the vehicle. If the trend is the same, the value is positive; otherwise, it is negative.

The trend eigenvector matrix Tr can be expressed as:

$$r = \begin{bmatrix} \Delta h_1 & \Delta h_2 & \ldots & \Delta h_{50} \\ \Delta v_1 & \Delta v_2 & \ldots & \Delta v_{50} \\ \Delta a_1 & \Delta a_2 & \ldots & \Delta a_{50} \\ \Delta z_1 & \Delta z_2 & \ldots & \Delta z_{50} \end{bmatrix} \tag{4}$$

We normalize the feature vectors by divining each row of the matrix Tr by the maximum trend change.

$$Tr' = \begin{bmatrix} \frac{\Delta h_1}{\Delta h_{max}} & \frac{\Delta h_2}{\Delta h_{max}} & \cdots & \frac{\Delta h_{50}}{\Delta h_{max}} \\ \frac{\Delta v_1}{\Delta v_{max}} & \frac{\Delta v_2}{\Delta v_{max}} & \cdots & \frac{\Delta v_{50}}{\Delta v_{max}} \\ \frac{\Delta a_1}{\Delta a_{max}} & \frac{\Delta a_2}{\Delta a_{max}} & \cdots & \frac{\Delta a_{50}}{\Delta a_{max}} \\ \frac{\Delta z_1}{\Delta z_{max}} & \frac{\Delta z_2}{\Delta z_{max}} & \cdots & \frac{\Delta z_{50}}{\Delta z_{max}} \end{bmatrix} \tag{5}$$

We tile the matrix Tr and obtain the final 1×200 trend eigenvector as the input.

$$Tr' = \begin{bmatrix} \Delta h'_1 & \ldots & \Delta h'_{50} & \Delta v'_1 & \ldots & \Delta v'_{50} & \Delta a'_1 & \ldots & \Delta a'_{50} & \Delta z'_1 & \ldots & \Delta z'_{50} \end{bmatrix} \tag{6}$$

The vector describes the trend of a behavioral feature. The larger the value of a feature, the more significant the feature is in this interval.

2.4 Construction of Temporal Feature Vector

The trend feature vector describes the behavioral features of the inertial navigation trajectory. However, in some cases, we need to consider that the factors of the trajectory change over time. Since the inertial navigation trajectory changes over time, only static features can be described by the feature vector. Therefore, it is necessary to construct a dynamic temporal feature vector for the input into the classification model. The temporal feature vector consists of attribute features that change over time in the inertial navigation trajectory. In a time sequence, the speed is a commonly used feature because it changes during turns and driving uphill and downhill. The acceleration is directly related to the speed. In addition, when vehicles turn, the turning radius varies with the size of the intersection and lane; thus, the heading angle changes. The elevation changes when the vehicle enters or exits a viaduct or ramps.

Therefore, we use the time stamp utc, velocity v, heading angle h, and elevation z to construct the temporal feature vector, which is defined as:

$$T_t = [utc_t, h_t, v_t, z_t] \tag{7}$$

where T_t represents the temporal feature vector consisting of the attribute elements at time t.

3 Hybrid Neural Network Extraction and Classification Model Based on CNN and GRU

A CNN is a feedforward neural network that uses convolution and has a deep structure; it is commonly used in deep learning [30, 31] and is capable of representation learning. The algorithm is shift-invariant and classifies the input information according to its hierarchical structure, learning the data features. Therefore, we selected the CNN as the classification model with the trend feature vector as the input data.

A gated recurrent unit (GRU) is a recurrent neural network that uses gate control. The network structure of the GRU is simpler than that of the LSTM network. The GRU combines the input gate and forget gate of the LSTM into one gate, called an update gate. Because of its unique structure, it is suitable for processing and predicting important events with very long intervals and delays in a time series. The GRU provides better performance, is easier to train, and has higher training efficiency than the LSTM. Since the inertial navigation trajectory data are time-series data, and the LSTM neural network is prone to overfitting with numerous parameters, the GRU layer is used as the hidden layer. The temporal feature vector is used as the input data of the model.

Since the two types of input vectors detect different features, the learning of different features may lead to different results. Therefore, we use the CNN and GRU network to learn the inertial navigation trajectory using the trend feature vector and temporal feature vector, respectively, as input data. Parameter weights are added in the final model training and optimization process. The final output represents a fusion of multiple behavioral features.

3.1 CNN for Trend Feature

(1) Input layer: we input the trend feature vector into the CNN network and extract the feature of each Tr.
(2) Convolution layer: three convolution layers are used according to the geometric, velocity, and elevation attributes. The output of each convolution layer is activated by a nonlinear function (Relu) as the input of the next convolution layer. No pooling layer is added between the convolution layers to avoid the loss of key information.

After convolution, a bias is typically added, and a nonlinear activation function is introduced. Here, the bias is defined as b, and the activation function is f_a. After adding the activation function, the result is as follows:

$$Q_i^k = f_a(W_k \times Tr)_i + b \tag{8}$$

(3) Fully connected layer: a fully connected layer is added at the end for the classification. The activation function is the Softmax function, and the output is the probability that the current trend feature vector is the behavioral feature (Fig. 5).

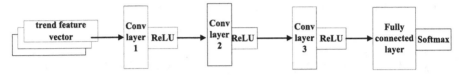

Fig. 5. The CNN structure

3.2 GRU for Temporal Feature

(1) Input layer: we input the temporal feature vector T_i into the GRU network.
(2) GRU layer: in the GRU network, there is no division between the internal state and external state in the LSTM network. Instead, it solves the problem of gradient disappearance and gradient explosion by adding a linear dependency between the current network state h_t and the last network state h_{t-1}.

The update gate is used to control the number of historical states h_{t-1} remaining in the output state h_t and the number of candidate states at the current time. The calculation of the renewal door is as follows:

$$z_t = \sigma(T_i^t W_{Tz} + h_{t-1} W_{hz} + b_z) \tag{9}$$

where T_i is the input vector of the t th time, i.e., the T components of the input sequence t. W_{Tz} and W_{hz} are the weight matrices of the current gate, and b_z is the bias value of the current neural network.

The output of the update gate is multiplied by the historical state h_{t-1} and the candidate state, respectively, and the sum is multiplied by $1-z_t$. Finally, the output of the network at the current time is:

$$h_{\bar{t}} = z_t \cdot h_{t-1} + (1 - z_t) \cdot \overline{h_t} \tag{10}$$

The function of the reset gate is to determine whether the candidate state at the current time depends on the network state of the previous time and the level of dependence. The network state h_t of the previous time is multiplied by the output r_t of the reset gate and used to calculate the candidate state at the current time. The calculation formula of the reset door is as follows:

$$r_t = \sigma(T_i^t W_{Tr} + h_{t-1} W_{hr} + b_r). \tag{11}$$

The value of r_t determines the dependence of the candidate state h_t on the state h_{t-1} of the previous time:

$$\bar{h_{\bar{t}}} = \tanh(T_i^t W_{Th} + (r_t \cdot h_{t-1})W_{hh} + b_h) \tag{12}$$

When the value of z_t is 0, and the value of r_t is 1, the update gate and reset gate in GRU network will no longer work, and the GRU network will degenerate into a simple cyclic neural network (Fig. 6).

(3)   Fully connected layer: a fully connected layer is added at the end for the classification. The activation function is a Softmax function, and the output is the probability that the current time-series feature vector is the behavioral feature.

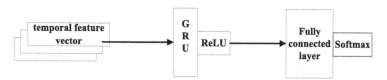

Fig. 6. The GRU structure

The overall structure is shown in Fig. 7.

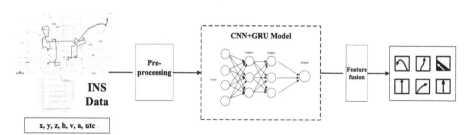

Fig. 7. The CNN + GRU structure

In this study, the input behavioral feature of the hybrid neural network is composed of the trend feature vector and temporal feature vector. Due to the different feature extraction rules, different classification results can be obtained for the same input features (e.g., a sharp turn has geometric features and speed changes). To solve this problem, we increase the weight at the end of the two neural networks. The matrix dot product of the weight and the output vector is used to combine the results of the two networks, and the behavioral feature with the highest probability is the final behavioral feature.

This approach is expressed as follows:

$$P = P_{Tr} \cdot W_{Tr} + P_{T_i} \cdot W_{T_i} \tag{13}$$

where P and W, respectively, represent the weight and probability of the classification results. It should be noted that the left turn, right turn, turn around, and other behavioral features depend on the trend features, and their values are greater than that of the trend features. Similarly, the uphill, downhill, acceleration, and deceleration features are temporal features in a given time range, and their weights are higher than those of the trend features to ensure that the output results are close to the real situation.

4   Experimental Results and Discussion

4.1   Experimental Environment

The computer used in the experiment is a core i7-10510u, 1.80 GHz CPU with 16 GB of memory and a Windows 10 operating system. Python is used to develop and implement the neural network. The learning framework is keras, and the back end is tensorflow.

The inertial navigation trajectory data was obtained in the Miyun area in Beijing in 2016. Each data record contains tracking information and the acquisition time.

In the experimental stage, all data are mixed; 70% of the data is selected as the training set, and 30% is the test set.

The proposed model is verified by comparing its extraction and classification accuracies with that of other models.

4.2   Model Evaluation

The accuracy, precision, recall, and F1 value are used to evaluate the model.

The parameters are defined as follows:

True positive (TP): the model correctly predicts the positive class.

True negative (TN): the model correctly predicts the negative class.

False positive (FP): the model predicts the negative class as a positive class (type I error).

False negative (FN): the model predicts the positive class as a negative class (type II error).

(1) Accuracy

Accuracy refers to the proportion of the number of correct samples to the total number of samples.

$$accuracy = \frac{TP + TN}{TP + TN + FP + FN} \tag{14}$$

(2) Precision

The precision is the ratio of the number of correct samples to the total number of samples (excluding the number of unclassified samples). This index is used to measure the accuracy of neural network feature extraction.

$$precision = \frac{TP}{TP + FP} \tag{15}$$

(3) Recall

The recall rate is the ratio of the number of samples correctly classified to the total number of samples that should be classified. It is used to measure the integrity of the experimental results.

$$recall = \frac{TP}{TP + FN} \tag{16}$$

(4) F1 value

The F1 value is the harmonic average of the recall rate and accuracy rate. Its value is close to the smallest value of precision and recall.

$$F1 = \frac{2 \times precision \times recall}{precsion + recall} \tag{17}$$

4.3 Experimental Results

(1) The Performance of the Proposed Model

The proposed hybrid neural network is used in five experiments, and the average value is used as the final result. The results of feature extraction are listed in Table 1.

Table 1. The results of the trajectory features extracted by the proposed model

Feature class	Sample	Correct	Wrong	Refuse	Accuracy
Left turn	100	94	6	0	94%
Right turn	100	92	8	0	92%
Turn around(L)	100	95	5	0	95%
Turn around(R)	0	0	0	0	0%
Left lane change	100	87	13	0	87%
Right lane change	100	89	11	0	89%
Uphill	100	92	8	0	92%
Downhill	100	92	8	0	92%
Accelerate	100	91	9	0	91%
Decelerate	100	91	9	0	91%
Total	900	823	77	0	–
Average	90	91.44%	8.56%	0	91.44%

The following can be concluded from the results in Table 2. (1) The proposed classification model has relatively higher accuracy for extracting geometric features, and the highest accuracy (95%) is obtained for turn around(L). The change in the heading angle is regular, which is conducive to the learning of the neural network. (2) Due to the temporal feature vector, the model considers the uphill, downhill, acceleration, and deceleration features. (3) Since lane changing behavior is similar to turning, especially significant lane changes due to different driving habits, the classification results are easily confused, and the accuracy is only 88%. (4) Due to the limitation of domestic traffic rules, there were few turn around(R) samples, which is 0 in this experiment, so it is no longer classified separately. (5) The average accuracy of the model is 91.44%.

(2) Comparison With Other Models

The performances of the proposed method and the DP, Support Vector Machine (SVM), CNN, and LSTM were compared; the results are shown in Table 2.

Table 2. Performance of several feature classification and extraction models

Model	Accuracy	Recall	F1
DP	79.98	78.76	78.78
SVM	86.90	86.63	86.66
CNN	90.11	89.82	89.84
LSTM	87.30	87.03	87.05
Our model	91.44	91.44	91.44

It is observed that the feature extraction and classification algorithms based on neural networks outperform traditional algorithm models regarding the accuracy, recall, and F1 value. Different driving behaviors under complex road conditions cannot be extracted if only the geometric features of trajectory data are considered. The experimental results show that integrating time-series features improves the feature extraction and classification accuracy. The proposed hybrid neural network has the advantages of considering geometric features, elevation, and velocity and provides higher accuracy than the other models.

5   Conclusion

Behavioral feature modeling and feature extraction and classification of vehicle trajectory data are key steps to achieve accurate positioning in navigation and positioning. High-accuracy feature matching can compensate for location uncertainty due to the lack of satellite signals. In this paper, inertial navigation trajectory data are analyzed using a CNN + GRU hybrid neural network model to extract and classify behavioral features. The experimental results show that the proposed model considers multiple behavioral features of the inertial navigation trajectory. The proposed model has higher extraction and classification accuracy than traditional models and mainstream neural network models and meets meet the requirements of behavioral feature extraction and classification.

Some limitations are observed for extracting similar behavioral, and the combination of multiple features is not considered. Therefore, our next study will focus on the following aspects: 1. In-depth analysis of the inertial navigation trajectory features; 2. Integration of the latest deep learning algorithm model; 3. Improvement of the model efficiency. The proposed model represents an improvement over previous models and achieves accurate and fast classification of the behavioral features of inertial navigation trajectory data, laying a solid foundation for subsequent matching calculation and error correction.

References

1. Zhao, L., Shi, G.: A method for simplifying ship trajectory based on improved Douglas-Peucker algorithm. Ocean Eng. **166**, 37–46 (2018)
2. Li, X., et al.: Feature extraction algorithm in consideration of the trend changing of track. J. Comput. Aided Des. Comput. Graph. **28**(8), 1341–1349 (2016)
3. Qiao, S.J., et al.: A trajectory feature extraction approach based on spatial coding technique. Sci. Sin. Inform. **47**, 1523–1537 (2017). https://doi.org/10.1360/N112017-00008
4. Wu, Q.Y., Wu, Z.F., Zhang, L.P.: A road geometric feature extraction method based on taxi trajectory data. CN108776727A (2018.11.09)
5. Zhu, L., et al.: Study on spatial-semantic trajectory based GPS track behavior signature detection. Comput. Appl. Softw. **31**(4), 72-74+87 (2014)
6. Zheng, Y.: Trajectory data mining: an overview. ACM Trans. Intell. Syst. Technol. **6**(3), 29 (2015)
7. Lu, C.W., et al.: Road learning extraction method based on vehicle trajectory data. Acta Geodaeticaet. Cartograph. Sin. **49**(6), 692–702 (2020)
8. Yu, J., et al.: Fine-grained abnormal driving behaviors detection and identification with smartphones. IEEE Trans. Mob. Comput. **16**(8), 2198–2212 (2017)
9. Stavros, G.C., Stratis, K., Alexander, C.: Learning driver braking behavior using smartphones, neural networks and the sliding correlation coefficient: road anomaly case study. IEEE Trans. Intell. Transport. Syst. **20**(1), 65–74 (2019)
10. Cui, S.M., et al.: A deep learning method for taxi destination prediction. Comput. Eng. Sci. **042**(001), 185–190 (2020)
11. Lv, J., et al.: T-CONV: A convolutional neural network for multi-scale taxi trajectory prediction. In: Proc of 2018 IEEE International Conference on Big Data and Smart Computing, pp. 82–89 (2018)
12. Sun, H., Chen, S.: Spatio-temporal trajectory prediction algorithm based on clustering based hidden Markov model. J. Chin. Comput. Syst. **40**(3), 472–476 (2019)
13. Ji, X.W., et al.: Intention recognition and trajectory prediction for vehicles using LSTM network. China J. Highway Transport **32**(6), 34–42 (2019)
14. Li, M.X., et al.: Predicting future locations with deep fuzzy-LSTM network. Acta Geodaeticaet. Cartograph. Sin. **47**(12), 1660–1669 (2018)
15. Wang, Z.S., Ye, Q.F., Long, W.: An automatic taxi target prediction algorithm based on artificial neural network. J. Anhui Vocation. Coll. Electron. Inform. Technol. **18**(1), 1–3 (2019)

16. Park, J., Cho, H.G.: Virtual running model for locating road intersections using GPS trajectory data. In: International Conference on Ubiquitous Information Management and Communication (2017)
17. Albanna, B.H., et al.: Semantic Trajectories: A Survey from Modeling to Application (2015)
18. Zhang, W., Li, S., Pan, G.: Mining the semantics of origin-destination flows using taxi traces. In: ACM Conference on Ubiquitous Computing, pp. 943–949. ACM, New York (2012)
19. Banharnsakun, A., Tanathong, S.: A hierarchical clustering of features approach for vehicle tracking in traffic environments. Int. J. Intell. Comput. Cybernet. 9(4), 354–368 (2016)
20. Zhan, X.Y., et al.: Urban link travel time estimation using large-scale taxi data with partial information. Transport. Res. C: Emerg. Technol. 33, 37–49 (2013)
21. Pan, G., et al.: Land-use classification using taxi GPS traces. IEEE Trans. Intell. Transport. Syst. 14(1), 113–123 (2013)
22. Lee, J.G., Han, J., Li, X.: trajectory outlier detection: a partition and-detect framework. In: International Conference on Data Engineering, pp. 140–149. IEEE Computer Society (2008)
23. Vries, G.K.D.D., Someren, M.V.: Machine learning for vessel trajectories using compression, alignments and domain knowledge. Expert Syst. Appl. 39(18), 13426–13439 (2012)
24. Lerin, P.M., Yamamoto, D., Takahashi, N.: Encoding travel traces by using road networks and routing algorithms. Intelligent Interactive Multimedia: Systems and Services, pp. 233–243. Springer, Heidelberg (2012)
25. Richter, K.F., Schmid, F., Laube, P.: Semantic trajectory compression: representing urban movement in a nutshell. J. Spatial Inform. Sci. 4, 3–30 (2012)
26. Song, H., et al.: Vehicle trajectory clustering based on 3D information via a coarse-to-fine strategy. Soft. Comput. 22(5), 1433–1444 (2018)
27. Aquino, A.L.L., et al.: Characterization of vehicle behavior with information theory. Eur. Phys. J. B 88(10), 257 (2015)
28. Zheng, Y, et al.: Learning transportation mode from raw GPS data for geographic applications on the web. In: Proceedings of the 17th International Conference on World Wide Web, pp. 247–256. ACM (2008)
29. Zhu, J., Jiang, N., Hu, B.: Application of multiple motion parameters of moving objects in trajectory classification. J. Earth Sci. 18(2), 143–150 (2016)
30. Goodfellow, I., Bengio, Y., Courville, A.: Deep Learning, vol. 1, 326–366. MIT Press, Cambridge (2016)
31. Gu, J., et al.: Recent advances in convolutional neural networks. arXiv Preprint arXiv:1512. 07108 (2015)

Optimal Siting of Rural Settlement Through a GIS-Based Assessment: A Case Study in China

Chuanhua Zhu[1,2(✉)], Dunhui Xiao[3], and QinYao Sun[1]

[1] School of Environment and Energy Engineering, Anhui Jianzhu University,
Hefei 230601, China
[2] Anhui Engineering and Technology Research Center of Smart City, Hefei 230601, China
[3] ZCCE, College of Engineering, Swansea University, Swansea SA1 8EN, UK

Abstract. The optimization of spatial allocation of rural settlements in China requires an evaluation of rural land suitability and location optimization for the convenience of rural residents living and working. The main idea behind this study was to propose a new methodology to locate the most suitable sites for rural residents as considering their potential demand and supply. A geographic information system (GIS) based multiple criteria decision-making (MCDM) model was used to identify the most suitable areas and the unsuitable parts. The weighted information value method was used to assigning weight for influential factors. A case study was conducted for Tangjiahui, a hilly area in the central of China, by performing a town-wide suitability assessment. The very high suitability area is 4621.05 hm^2, where is close to the road and with a moderate elevation and low slope, which contains 70.29% of residents. Finally, the maximal covering model was performed to determine the most suitable locations. Six sites were selected from 312 potential sites, which cover whole potential demand settlements with an average travel distance of 2.9 km. The result of location optimization is reasonable and therefore the methodology is applicable.

Keywords: Hilly area · Rural settlement · Weighted information value model · Land suitability analysis · Location optimization · Maximal covering

1 Introduction

In recent years, land use planning of rural areas are becoming popular because of the rapid expansion of urban area leading the shortage of farming land in China. As a consequence, many studies have analyzed the relationship between the spatial distribution of rural settlement and related influencing factors, including the natural environment, economic development and social culture, on the basis of location theory [1, 2], to determine the land suitability and relocate famers' houses. While western studies have focused on the pattern evolution of rural settlements, the landscape evolution of rural settlements, and the relationship between rural industries and rural settlements [3–5]. China is a country has spacious territory, rural settlements are located in various type of land, such as hilly, plain and delta. However, in order to utilize the potential land,

© Springer Nature Switzerland AG 2021
G. Pan et al. (Eds.): SpatialDI 2021, LNCS 12753, pp. 57–71, 2021.
https://doi.org/10.1007/978-3-030-85462-1_5

whatever the types of rural residential areas, the land suitability evaluation is first carried out, and then the optimization analysis. In the wider domain of geo-computation and spatial analysis, adopting the Multiple criteria decision-making [6, 7] approach for land suitability analyses and a number of different methods are well-established for weight of criteria. The most popular method for site selection is the Analytic Hierarchy Process (AHP), Delphi and their combination with other methods [8–13], such as combination of Delphi and weighted entropy, combination of the AHP and weighted entropy, which are all objective that depend on the expert opinion in some extent. In addition, they produced too many potential sitings for decision-makers to choose from. Optimization analysis generally also use the qualitative methods, which just give the suggestion from macro view without the blueprint for resettlement [10, 11].

However, limited research has been conducted using quantitative methods. Xie and Fan used the weighted Voronoi tessellation to represent the influenced area of rural settlement by using the comprehensive index value as the weight value [12, 13], so as to determine the relocation and retention of settlement. But the analysis of Vorinoi tessellation is based on Euclidean distance and does not take the actual demand of settlements and capacity of supply into account. Maximal covering model is a kind of location-allocation model, which considers not only the needs of demand points but also the capacity of supply, which had been successfully applied in the location optimization of police patrol areas, alternative-fuel stations, school, biogas plants, health care organizations, bioenergy facilities and bike-sharing station [14–22]. This model will be introduced in our study to analyze the location optimization of rural settlement.

The overall objective of this study work is to develop a methodology to allocate the optimal settlement site for rural residents considering their demand and capacity of candidate sites based on real road networks. The specific objectives include: (1) Developing a framework to assess the suitable location for rural settlement using a set of environmental factors; (2) Integrate GIS and weighted information value method for the development of the framework; (3) Implement a location-allocation analysis to select the most suitable sites for rural settlements in the study area.

This paper has four sections. Section 1 proceeds with a review of land suitability analyses that have been used for rural settlements and applications of location-allocation algorithms in this context. Before introducing the specific case study used to demonstrate this methodology (Sect. 2), Sect. 3 provides a detailed description of the proposed weighted information value method and maximal covering model. The results and discussion are presented in Sect. 4.

2 Methodology

There are two stages of the proposed method, that is land suitability analysis and location optimization, for identifying optimal sites for settlements. First, it is necessary to know the distribution of the potential demand using land suitability analysis, also known as Multiple Criteria Decision-Making, which use different types of spatial analyze tools such as reclassify and map algebra to determine the potential supply and demand locations, and then location optimization, which employ search heuristics to satisfy location-allocation problem such as the maximal covering location problem and seek to match spatially distributed supply and demand.

2.1 Land Suitability Analysis

In this work, a land suitability analysis was performed to identify suitable and unsuitable areas. In order to do so, some environmental and social factors were considered. According to the references [8, 10–12] and data availability, several influence factors have been selected for land use suitability analysis of rural settlements, those are slope, aspect, elevation, distance to river, distance to road, distance to town and central villages. Each influence factors, the continuous variable, are reclassified into discrete variable by quantifying the influence on the suitability, such as the slope is divided into five categories, which are 0°–5°, 5°–10°, 8°–15°, 15°–22.5° and 22.5° or more; the aspect is divided into five categories, which are south, flat, southeast, southwest, east, west, northeast, northwest and north; the elevation is divided into five categories, which are 0–200 m, 200–400 m, 400–600 m, 600–800 m and 800 m or more; the distance to river is divided into five categories, which are 0–100 m, 100–200 m, 200–400 m, 400–600 m and 600 m or more; the distance to road is divided into five categories, which are 0–100 m, 100–200 m, 200–500 m, 500–1000 m and 1000 m or more; the distance to market town and central villages is divided into five categories, which are 0–200 m, 200–500 m, 500–1000 m, 1000–2000 m and 2000 m or more.

The weighted information value method was used to combine all these raster maps to produce the final suitability map. The method proposed by Van Westen is also called the statistical index method, was employed in landslide susceptibility mapping under a GIS environment [23]. In this method, a weighted value for a parameter class is defined as the natural logarithm of the rural settlement density class, divided by the settlement density in the entire map. The formula was given below that forms the basis of the approach:

$$Wi = \ln \frac{Densclass}{Densmap} = \ln \frac{\frac{Npix(Si)}{Npix(Ni)}}{\frac{SNpix(Si)}{SNpix(Ni)}} \tag{1}$$

Where

W_i = The weight given to a certain parameter class;
A $Densclass$ = Settlements density within the parameter class;
$Densmap$ = Settlements density within the entire map;
$Npix(Si)$ = Number of pixels that contain settlements in a certain parameter class;
$Npix(Ni)$ = Total number of pixels in a certain parameter class;
$SNpix(Si)$ = Number of pixels that contain settlements in the entire map;
$SNpix(Ni)$ = Total number of pixels in the entire map.

The W_i method is based on statistical correlation of the settlements inventory map with attributes of different parameter maps. In this study, every parameter map was crossed with the existed settlements map, and the density of the settlements in each class was calculated. Correlation results were stored in resultant raster and the density of the settlements per parameter class was calculated. Then the W_i value of each attribute was calculated. Finally, all layers were summed up and a resultant suitability map was obtained. The final suitability map was divided into five classes by the Natural Break method. The classes are very low suitability, low suitability, moderate suitability, high suitability and very high suitability.

2.2 Location-Allocation Model

Maximal covering model can be applied to the problem of generating optimal concentrated-settlements with the following formula:

$$Maximize\ Z = \sum_{i\in I} a_i y_i \tag{2}$$

Subject to:

$$\sum_{j\in N_i} x_j \geq y_i\ for\ all\ i \in I \tag{3}$$

$$\sum_{j\in J} x_j = P \tag{4}$$

$$x_j = (0, 1)\ for\ all\ j \in J \tag{5}$$

$$y_i = (0, 1)\ for\ all\ i \in I \tag{6}$$

$$\sum_{i\in N_j} a_i x_j \leq M_j\ for\ all\ j \in J \tag{7}$$

Where:

Z = the total demand of rural residents serviced.
I = the set of known settlements locations.
J = the set of potential locations for concentrated-settlements.
$x_j = 1$ if a concentrated-settlements covers settlements locations j, and 0 otherwise.
$y_i = 1$ if an settlement locations at i is covered by at least one l concentrated-settlement, and 0 otherwise.
P = the number of concentrated-settlements to be located

$$N_i = \{j \in J | d_{ij} \leq S\}$$

d_{ij} = the shortest distance from settlement location i to concentrated-settlement location j.

$$N_j = \{i \in I | d_{ij} \leq S\}.$$

S = the acceptable service distance or time.
a_i = the demand of settlements location i.
M_j = the maximum capacity that each concentrated-settlement at j can serve.

In this model, it is assumed that the daily work of rural residents needs to go back and forth between the work zone and the existing settlements, and the distance between the existing settlements and the work zone is negligible, that is the rural residents need to go back and forth between the existing settlements and the concentrated-settlements area after location optimization.

3 Case Study

3.1 Study Site

The proposed method has been applied in Tangjiahui, a town of Jinzhai county in China. Jinzhai is located in the hinterland of Dabie Mountain area where is located in the centre China. And Tangjiahui is in the west part of Jinzhai and has 11 administrative villages, 1 street community, and 306 natural village, with a total area of 26850 hectares and more than 50000 inhabitants. The study area is characterized by hilly area that is about 80.4% and the forest coverage is 65.2%. Because of the natural terrain and environment, rural settlements are scattered in the more flat and less elevation area but more difficult to the road, which is the main reason for vulnerable people difficult to access the public facilities, as were shown in Fig. 1.

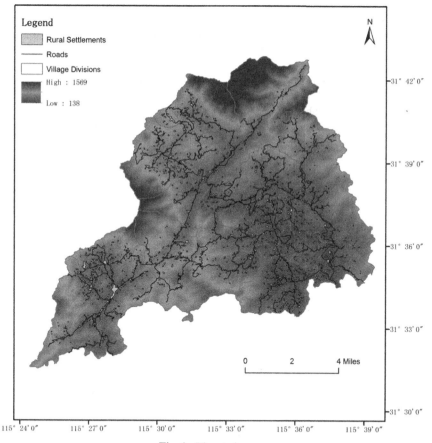

Fig. 1. The study area

3.2 Data Used

The data used in this study involves: digitized data interpreted from Google images, DEM data, land use planning, demographic and socioeconomic data. The digitized data includes rural residential area, road, river, centers (governmental town and village). The DEM dataset, with a grid resolution of 30 m, is provided by Geospatial Data Cloud site, Computer Network Information Center, Chinese Academy of Sciences (http://www.gsc loud.cn). The slope and aspect data are derived from the DEM data via spatial analysis, in particular, the surface analysis method. The land use planning, demographic and socioeconomic data is obtained from local governmental information website (http://www.ahjinzhai.gov.cn).

3.3 Evaluation of Single Data Layer

In ArcGIS, Euclidean distance and reclassify tools are used to process each data layer, and the derived data layer interact with the settlements layer (binary raster layer) by using zonal statistics tool to calculate the information value W_i, as shown in Table 1.

Table 1. Influence factors for suitability analysis of rural settlements

Data layers	Classes number	Classes	Class area (%)	Settlements areas (%)	W_i
Slope	1	0°–5°	11.07	25.52	0.835
	2	5°–10°	23.40	56.48	0.881
	3	10°–15°	22.06	19.44	−0.127
	4	15°–22.5°	24.32	9.08	−0.985
	5	>22.5°	19.15	3.12	−1.814
Aspect	1	South, Flat	16.15	20.72	0.249
	2	Southeast, Southwest	29.06	34.03	**0.158**
	3	East, West	21.80	20.13	−0.079
	4	Northeast, Northwest	21.68	17.44	−0.217
	5	North	11.30	7.68	−0.387
Elevation	1	0–200 m	2.66	5.92	0.799
	2	200–400 m	17.00	30.96	0.600
	3	400–600 m	20.96	31.85	0.418

(continued)

Table 1. (*continued*)

Data layers	Classes number	Classes	Class area (%)	Settlements areas (%)	W_i
	4	600–800 m	20.83	17.66	−0.165
	5	>800 m	38.55	13.61	−1.041
Distance to river	1	0–100 m	12.00	22.33	0.621
	2	100–200 m	8.18	15.93	0.667
	3	200–400 m	15.37	17.16	0.110
	4	400–600 m	13.78	14.05	0.019
	5	>600 m	50.67	30.54	−0.506
Distance to road	1	0–100 m	21.50	87.41	1.403
	2	100–200 m	15.43	11.10	−0.329
	3	200–500 m	11.41	1.17	−2.275
	4	500–1000 m	7.27	0.27	−3.283
	5	>1000 m	44.39	0.04	−6.989
Distance to center	1	0–200 m	0.63	5.11	2.101
	2	200–500 m	3.38	9.57	1.042
	3	500–1000 m	11.33	21.19	0.626
	4	1000–2000 m	31.73	42.01	0.280
	5	>2000 m	52.93	22.11	−0.873

The W_i calculated by weighted information value method indicate the correlation of settlements with data layer, as shown in Table 1 and Fig. 2 that the parameter layers with strong positive correlation with the settlements layer are as followed: 0°–5° and 5°–10° of slope class, south and flat class of aspect, 0–200 m and 200–300 m of elevation class, 0–100 m and 100–200 m class of distance to river, 0–50 m class of distance to road and 0–200 m and 200–500 m class of distance to center.

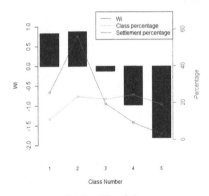

a. Reclassify of slope

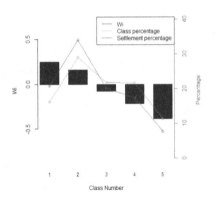

b. Reclassify of aspect

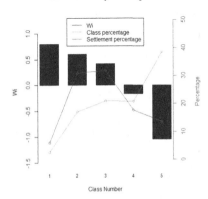

c. Reclassify of elevation

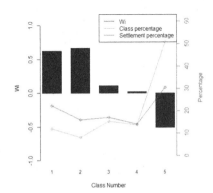

d. Reclassify of distance to river

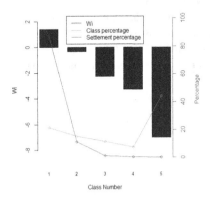

e. Reclassify of distance to road

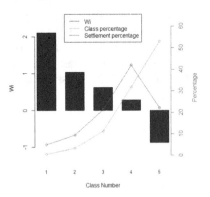

f. Reclassify of distance to town and village center

Fig. 2. Reclassify of influence factors of rural residential land in Tangjiahui

4   Result and Discussion

4.1   Suitable Areas Suggestions

The suitability index map is derived from the influence layers. Add operator in the map algebra toolset was used to sum up all data layers, then the resultant suitability map was divided to five classes according to the information value stored in each single grid, as shown in Fig. 3.

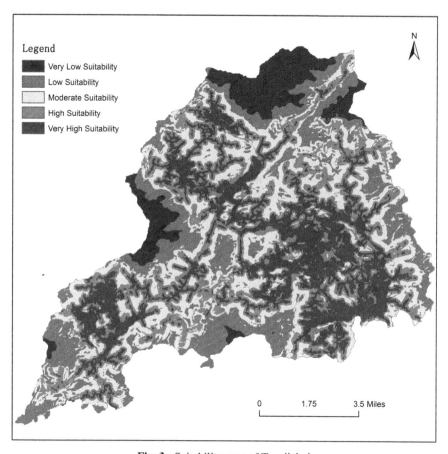

Fig. 3. Suitability map of Tangjiahui

4.2   Validation of the Land Suitability Model

The zonal statistics tool was use to statistic the distribution of settlements layer in suitability classes. As shown in Table 2, the very high suitability area is only accounts for 18.03% of the total area, but contains about 70.29% of the existing settlements, and the b/a ratio of classes from unsuitable area to suitable area is gradually increased, which indicate the classification is good and model is useful.

Table 2. Model verification of suitability zoning

Suitability level	Area (hm^2)	a (%)	b (%)	c (%)	b/a
Very low suitability	7237.35	28.24	0.04	0.004	0.001
Low suitability	4417.56	17.24	0.03	0.004	0.002
Moderate suitability	4068.9	15.88	3.10	0.50	0.20
High suitability	5283.81	20.62	26.53	3.31	1.29
Very high suitability	4621.05	18.03	70.29	10.04	3.90

Notation: a indicated the grid number of division as a percentage of the total grid number; b indicated the grid number of settlements in division as a percentage of the total grid number of settlements; c indicated the grid number of settlements in division as a percentage of the grid number of division.

4.3 Location Optimization

There are three steps to decide the optimal sites. Firstly, determine the potential demand (settlements located in unsuitability) and supply (concentrated-settlements). The grids in the suitability map with high suitability or above were converted to vector format,

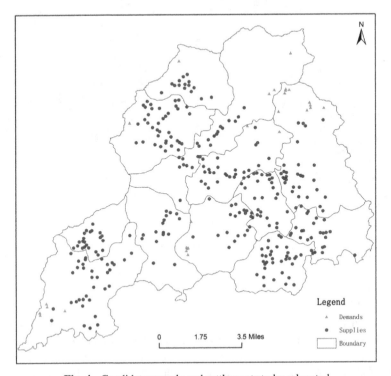

Fig. 4. Candidate parcels and settlements to be relocated

and then intersect with the construction areas that allowed by government, therefore the candidate parcels with the area more than 1000 m^2 were selected, which were used as potential supply. The raster grids with low suitability or below were also converted to vector format, and then intersect with the existing settlements, therefore the selected settlements were used as potential demand. Secondly, all the potential demand and supply parcels were converted to points and then connected to the nearest node in the road network, which were set as the demand point and supply point respectively. Finally, location-allocation model of network analyst module in ArcGIS is used to solve the capacitated maximum covering problem with constrains.

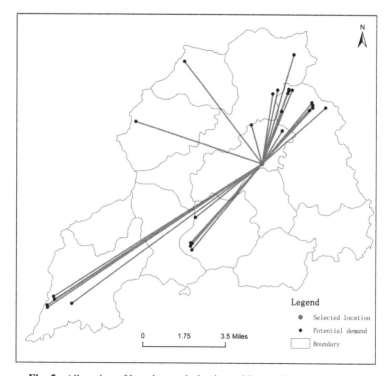

Fig. 5. Allocation of location optimization subject to P = 1, S = 50000

Figure 4 shows there are 312 parcels of potential supply, of which the largest capacity is 965 (counted by 5 residents per 120 m^2), and 28 parcels of potential demand include 662 residents.

Maximal covering models with capacity and without capacity were both used to identify optimal sites for concentrated-settlements. Indices include average travel distance, the longest single distance, the least single distance, the ratio of covering and the ratio of service were used to evaluate the model. Table 3 shows when choosing the uncapacitated maximal covering without capacity and just 1 candidate site to be chosen, the location-allocation map is show in Fig. 5, that is the covering rate is 100%, but the single longest distance is 29.9 km and average travel distance is 14.2 km, of which result

Table 3. The statistics indices of location optimization

The number of located sites (P)	ID of selected sites	The acceptable service distance S in meters	Is capacitated or not	The number of coverage settlements/the ratio of coverage (%)	The number of people serviced/the ratio of service (%)	Average travel distance in meters	Least single distance in meters	Longest single distance (m)
1	414	50000	not	28/100	662/100	14220	4153	29971
1	null	10000	yes	14/50	399/60.3	null	null	null
2	null	10000	yes	21/75	552/83.4	null	null	null
3	null	10000	yes	25/89.3	599/90.5	null	null	null
4	null	10000	yes	27/96.4	658/99.4	null	null	null
5	null	10000	yes	27/96.4	658/99.4	null	null	null
6	318,337,354, 356,523,621	10000	yes	28/100	662/100	2910	1155	9551

is unreasonable; while when choosing the model with capacity and 6 candidate site to be chosen with the acceptable distance of 10000 m, the location-allocation map is show in Fig. 6, and covering rate is 100%, and the single longest distance is 9.5 km and average travel distance is 2.9 km, of which result is reasonable.

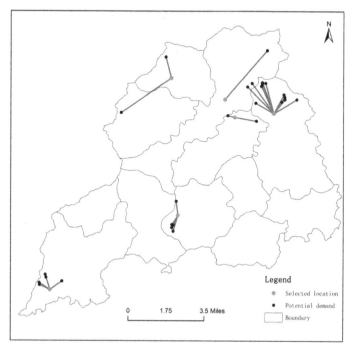

Fig. 6. Allocation of location optimization subject to P = 6, S = 10000

5 Conclusion

In this study, we have shown the possibility of the maximal covering model integrated with the land suitability evaluation model and applied to a proposal for the optimal location of rural concentrated-settlements in Tangjiahui, a town in Dabie mountain area of China. The result of land suitability evaluation showed the area of very high suitability is 4621.05 hm^2, which is close to the road and located in zone with the moderate elevation and low slope, with the 70.29% of residents. 6 sites were selected from 312 potential sites by capacitated maximal covering model, which cover the whole potential demand settlements with an average travel distance of 2.9 km, therefore result of optimal sitting is reasonable. The results produced by the proposed method are useful and applicable for local administration in reallocating the rural settlements. Nevertheless, it is clear that there are some limitations. Firstly, the influence factors of land suitability evaluation need to be considered thoroughly, which the distribution of rural settlements is not only influenced by the natural environment but also by the social, economic, ethical, and other facets. The assignment of settlements demand is the second limitation, which was distributed by the area of settlements rather than the actual data because of the difficulty of the collection at the village scale. In conclusion, the methodology proposed in this paper considers the actual demand and supply capacity, which is a practicable and quantifiable way to determine the optimal location for rural settlements relocating.

Acknowledgement. This study was carried out with the financial support of College Students Innovation and Entrepreneurship Training Project of Anhui Jianzhu University (Grant No. 2019051). Dr. Xiao would like to acknowledge the support of EPSRC grant PURIFY (EP/V000756/1) and the Royal Society International Exchanges 2019 Cost Share (IEC\NSFC\191037).

References

1. Li, J., Li, X.: A review on location of the rural settlements. Hum. Geogr. **23**(4), 23–27 (2008)
2. Jiang, G., Zhang, F., Qin, J., Zhang, L., Gong, P.: Relationship between distribution changes of rural residential land and environment in mountainous areas of Beijing. Trans. CSAE **22**(11), 85–92 (2006)
3. Carrion-Flores, C., Irwin, E.G.: Determinants of residential land use conversion and sprawl at the rural-urban fringe. Am. J. Agr. Econ. **86**(4), 889–904 (2004)
4. Paquette, S., Domon, G.: Changing ruralities, changing landscapes: exploring social recomposition using a multi-scale approach. J. Rural. Stud. **19**(3), 425–444 (2003)
5. Sevenant, M., Antrop, M.: Settlement models, land use and visibility in rural landscapes: two case studies in Greece. Landsc. Urban Plan. **80**(4), 362–374 (2007)
6. Pereira, J.M.C., Duckstein, L.: A multiple criteria decision-making approach to GIS-based land suitability evaluation. Int. J. Geogr. Inform. Syst. **7**(5), 407–424 (1993)
7. Kabak, M., Erbas, M., Cetinkaya, C., Ozceylan, E.: A GIS-based MCDM approach for the evaluation of bike-share stations. J. Clean. Prod. **201**, 49–60 (2018)
8. Gao, H., Zhang, C., Cai, G., Luo, X.: Land suitability evaluation of rural settlements in Krast Mountains based on GIS. Res. Soil Water Conserv. **22**(2), 200–203 (2015)
9. Mohib-Ul-Haque Khan, M., Vaezi, M., Kumar, A.: Optimal siting of solid waste-to-value-added facilities through a GIS-based assessment. Sci. Total Environ. **610–611**, 1065–1075 (2018)
10. Qu, Y., Zhang, F., Jiang, G., Guan, X., Guo, L.: Suitability evaluation and subarea control and regulation of rural residential land based on niche. Trans. CSAE **26**(11), 290–296 (2011)
11. Jiang, L., Lei, G., Zhang, J., Zhang, Y., Li, J.: Analysis of spatial distribution and optimization of rural settlement. Res. Soil Water Conserv. **20**(1), 224–229 (2013)
12. Xie, Z., Zhao, R., Jiang, P., Liu, H., Zhu, W.: The rural residential space reconstruction in Loess Hilly Regions: a case study of Yuzhong county in Lanzhou. Geogr. Res. **33**(5), 937–947 (2014)
13. Fan, T., Yang, Q., He, J., Pan, F., Deng, Y.: Spatial distribution optimization of rural residential land in hilly areas: a case study of Haitang town in Changshou district. Geogr. Res. **34**(5), 883–894 (2015)
14. Curtin, K.M., Hayslett-McCall, K., Qiu, F.: Determining optimal police patrol areas with maximal covering and backup covering location models. Netw. Spat. Econ. **10**, 125–145 (2010)
15. Christopher, U., Michael, K., Seow, L.: A model for location of capacitated alternative–fuel stations. Geogr. Anal. **41**, 85–106 (2009)

16. Kong, Y., Wang, Z.: Optimal location-allocation for county-level compulsory school site selection using GIS and integer linear programming. J. Geo-inform. Sci. (Chin. Ed.) **14**(3), 299–304 (2012)
17. Qie, H., Li, Y.: GIS-based spatial distribution characteristics of settlements and its optimization in the tropical Lingshui County of Hainan Province. Sci. Soil Water Conserv. **15**(5), 78–85 (2017)
18. Sahoo, K., Mani, S., Das, L., Bettinger, P.: GIS-based assessment of sustainable crop residues for optimal siting of biogas plants. Biomass Bioenerg. **110**, 63–74 (2018)
19. Syam, S.S., Côté, M.J.: A location-allocation model for service providers with application to not-for-profit health care organizations. Omega **38**, 157–166 (2010)
20. Comber, A., Dickie, J., Jarvis, C., Phillips, M., Tansey, K.: Locating bioenergy facilities using a modified GIS-based location-allocation-algorithm: considering the spatial distribution of resource supply. Appl. Energy **154**, 309–316 (2015)
21. Sultana, A., Kumar, A.: Optimal siting and size of bioenergy facilities using geographic information system. Appl. Energy **94**, 192–201 (2012)
22. Garcia-Palomares, J.C., Gutierrez, J., Latorre, M.: Optimal the location of stations in bike-sharing programs: a GIS approach. Appl. Geogr. **35**, 235–246 (2012)
23. Cevik, E., Topal, T.: GIS-based landslide susceptibility mapping for a problematic segment of the natural gas pipeline, Hendek (Turkey). Environ. Geol. **44**, 949–962 (2003)

Spatial Explicit Evaluation of Land Use Sustainability Based on Grid analysis—A Case Study of the Bohai Rim

Yu Pang[1], Hongrun Ju[1(✉)], and Shunkang Lu[2]

[1] School of Tourism and Geography Science, Qingdao University, Qingdao 266071, China
juhr@qdu.edu.cn
[2] Basic Teaching Center, Ocean University of China, Qingdao 266100, China

Abstract. Identifying the land use sustainability at an explicit spatial level in large areas is critical for optimizing the land use. The evaluation of sustainable land use generally takes the administrative units at the provincial, prefecture, or county level as the spatial scale, which ignore the spatial variation within the administrative units. This paper aims to put forward a novel evaluation method on sustainable land use evaluation and evaluate the land use sustainability at a grid level of 5 km in Bohai Rim of China. This paper first established an index system of sustainable land use evaluation frame with 13 indicators in four criterion levels: ecology, economy, society, and spatial pattern. Then, the 13 indicators were expressed at 5 km grids level. At last, the land use sustainability was evaluated of each grids based on the spatial principal component analysis. The results showed that the ecological environment in the Bohai Rim has the greatest impact on the level of sustainable land use. The high level of sustainable land use is generally distributed in the northwest and northeast in Bohai Rim, and the low level of sustainability is concentrated in the southwest of the Bohai Rim. This paper proposed a novel method to evaluate of the land use sustainability at a spatial explicit level. The method could offer more spatial information of sustainable land use to help government propose more efficient policies regarding improving land use sustainability.

Keywords: Spatial explicit evaluation · Land use sustainability · Spatial principal component analysis · Bohai rim

1 Introduction

Since the 1990s, sustainable development research has been vigorously developed in response to the shortage of resources, the deterioration of the ecological environment, and rapid urbanization. Among the 17 Sustainable Development Goals (SDGs) proposed by the United Nations, seven of the goals are related to land use [1]. It can be seen that sustainable land use, as an important part of sustainable development research, has attracted much attention from international organizations and countries. The Sustainable Land Use Evaluation Outline (FESLM) promulgated by the United Nations Agriculture

© Springer Nature Switzerland AG 2021
G. Pan et al. (Eds.): SpatialDI 2021, LNCS 12753, pp. 72–82, 2021.
https://doi.org/10.1007/978-3-030-85462-1_6

and Food Organization (FAO) defines sustainable land management as based on the existing land use system that enables land users to maintain or enhance the ecological support functions of land resources while maximizing the economic and social benefits of land use [2]. Land sustainable use evaluation can provide data supporting the rational and sustainable use of regional land and is helpful to guide the formulation of regional land management policies. The core of land sustainable use evaluation is the construction of evaluation model, including the research of index system framework, index selection, index weight determination.

The research on the index system framework is the foundation and the core of the land sustainable use evaluation model. The following three frameworks are commonly used in existing research: (1) "Productivity-Security-Protection-Viability-Acceptability" framework [2, 3]; (2) "Economy-society-ecology" framework [4, 5]; (3) "Pressure-state- response" (PSR) framework [6, 7]. First, the "Productivity-Security-Protection-Viability-Acceptability" framework is determined by the FESLM, which specifically refers to the five evaluation criteria for sustainable land use: land productivity, land security, protection of water and soil resources, economic feasibility and social acceptability. This framework lays the foundation for the development of a regional sustainable land use evaluation index system. Under this framework, Chen constructed an evaluation index system for sustainable land use, which is composed of 28 indicators in the five standard levels of "production, protection, stability, economic vitality and acceptability" [8]. At the same time, according to the five-evaluation standard framework proposed by FESLM, scholars have improved and simplified the framework, and formed "Economy-Society- Ecology" evaluation index system framework, which has been widely used in practiced. For example, Peng et al. summarized the goal of sustainable land use as three aspects: the ecological rationality, the economic effectiveness and the social acceptability [5]; Chen et al. proposed a index system from ecological, economic and social aspects, and the evaluation method and comprehensive integration method for each index were developed [9]. In addition, the PSR evaluation index system framework reflects the interaction between humans and the environment. Based on the PSR model, Xie et al. constructed a land use sustainability evaluation index system and comprehensive evaluation model composed of 14 indicators and evaluated the Poyang Lake ecological economic zone [6].

The determination of index weight and comprehensive evaluation are usually complicated, which are the emphases of land sustainable use evaluation model. Entropy method [10], principal component analysis method [11], analytic hierarchy process (AHP) and other methods have been used in existing studies to conduct comprehensive evaluation of multiple indicators. AHP mainly determines the weight of each index based on expert knowledge [12]. AHP is a subjective weighting method, and it has a relative low stability. Entropy method and principal component analysis method use mathematical statistics. They assigned weights based on the numerical characteristics of the indicators, which are objective weighting methods and have good norms. However, when the index is integrated, it faces the problem of matching the index in grids and the administrative unit. In the indicator system of sustainable land use, resources, environment, ecology, and other information are mostly spatially different, while economic and social statistical information is more time-dynamic [9]. When using the above methods for comprehensive evaluation of weights and indicators, the grid data is usually converted into administrative unit data to solve the spatial matching problem between them. Therefore, the evaluation of sustainable land use generally takes the administrative unit at the provincial, prefecture, or county level as the spatial scale.

The existing land sustainable use evaluation studies have provided a good foundation for the formulation of regional land use management, sustainable development, and other policies. However, two points could be improved to better evaluation the land use sustainability. First, the sustainability of land use includes not only the continuous use of land use in time, but also the optimization of spatial pattern [13]. The traditional land sustainable use evaluation system based on the framework of "Productivity-Security-Protection-Viability-Acceptability", "Economy-Society-Ecology" and "Pressure -State-response" focuses on the evaluation of land function and lacks the research on the impact of land use spatial pattern on land sustainable performance. Certain spatial pattern of land use can realize certain land functions. Therefore, it is necessary to increase the evaluation of the spatial pattern of land, urban structure, and other spatial aspects on the sustainable development process. Secondly, research on quantitative spatial evaluation of sustainable land use at a more spatial explicit scale is insufficient. The sustainable land use evaluation results are commonly a single numerical indicator within an administrative division unit, which is difficult to reflect the spatial heterogeneity of the land sustainable use level within the administrative regions.

This paper aims to (1) improve the framework of current evaluation index by adding land use spatial pattern index into the analysis; and (2) realize more explicit spatial evaluation of sustainable land use at the grid scale with the technology of GIS and land use data. This paper could offer useful spatial information regarding sustainable land use planning and guide the spatial optimize the land use.

2 Study Area, Data and Methods

2.1 Study Area

The Bohai Rim region is a "C-shaped" region composed of the Liaodong Peninsula, the Shandong Peninsula and the North China Plain, including Beijing, Tianjin, Hebei, Shandong, and Liaoning. The area is a typical example of a rapid urbanizing area in China. The urbanization rate of the Bohai Rim region rose rapidly from 54.73% in 2010 to 59% in 2015 [14, 15], becoming one of the fastest growing regions in China during this period. The economic growth of this area is also leading in China, with an increase in GDP from 12 trillion RMB in 2010 to 16.37 trillion RMB in 2015. The total population (246 million) and land area (523,429 km^2) in this region represent 18.3% and 10.95% of the national total number, respectively [16]. At the same time, land exploitation and utilization level is high in Bohai Rim. The used land area accounts for more than 84% of the total area, which was higher than the national average [17]. Therefore, it is necessary and representative to carry out spatial evaluation quantitative of sustainable land use of Bohai Rim in China.

2.2 Data Sources

This article uses three kinds of data: (1) Land use raster data with a spatial resolution of 100 m in 2010, from the 1:100,000 remote sensing monitoring database for land use status in China constructed by the Chinese Academy of Sciences. A hierarchical land use classification system with six primacy level types and 25 secondary level types was adopted. The first level types included cropland, woodland, grassland, waterbodies, built-up land and unused land. (2) The social, economic and ecological data include the investment in fixed assets, the income of urban residents, the income of rural residents, and fertilizer use amount from 2011 "Beijing Statistical Yearbook", "Tianjin Statistical Yearbook", "Liaoning Statistical Yearbook", "Hebei Economic Yearbook", "Shandong Statistical Yearbook". The data of GDP and population are from Dataset of the spatial distribution of GDP and population in China at 1 km^2 grid [18]. (3) The river vector data from the national basic geographic information system.

2.3 Methods

Evaluation Framework and Choice of Indicators. Based on the traditional "Economy-Society-Ecology" framework of sustainable land use evaluation, this paper proposed an "Economy-Society-Ecology-Pattern" Four Dimensions framework by addressing the spatial pattern of land as an important aspect that influencing the sustainability of land use. Then, we adopted the following three requirements when choosing evaluation indicators: (1) the indicators should have a direct or indirect connection with the sustainability of land use; (2) the indicators can reflect the diverse aspects of economy, society, ecology and spatial pattern; (3) data for indicator calculation were available. Finally, thirteen indicators were selected based on the four dimensions framework, three requirements and literature reviews (Table 1).

Table 1. The indicators for evaluation of sustainable land use in the Bohai Rim

Goal	Criteria	Indicators	Abbreviation
Sustainable land use	Economy	GDP	GDP
		Income of urban residents	UPI
		Income of rural residents	RPI
		Investment in fixed assets	IFA
	Society	Land urbanization	LU
		Population density	PD
		Per capita cultivated land area	PCCLA
		Per unit area grain yield	PUAGY
	Ecology	Forest coverage rate	FCR
		Fertilizer used per area	FUPA
		Distance from water	DFW
	Pattern	Shannon's evenness index	SHEI
		Contagion index	CONTAG

Spatial Discretization of the Indicators. To ensure reasonable results from evaluation, we transformed the indicators of income of urban residents, income of rural residents, investment in fixed assets, grain yield and fertilizer used into a grid size of 5 km^2 based on land use data on which the indicator's activity occurs. First, we assumed that each indicator per land area was uniform in a prefectural unit and then we discretized the indicators into 5 km^2 grids based on Eq. (1):

$$INDIC_i = INDIC_j/A_j \times A_i \tag{1}$$

where $INDIC_i$ is the indicator value in grid i; $INDIC_j$ is the indicator value of the prefectural unit j; A_j is the area of corresponding land use type in the prefectural unit j; and A_i is the area of corresponding land use type in grid i. The corresponding land use type of income of urban residents is urban land, the corresponding land use type of income of rural residents is rural settlements, the corresponding land use type of investment in fixed assets is industrial land, and the corresponding land use type of grain yield and fertilizer used is cropland.

Spatial Principal Component Analysis. Principal component analysis (PCA) involves a mathematical procedure that transforms a number of (possibly) correlated variables into a (smaller) number of uncorrelated variables called principal components, which are linear combinations of the original variables. The main objective of the PCA is to discover or to reduce the dimensionality of the data set and to identify new meaningful underlying various. The SPCA application assists and guides a user in doing PCA in a spatial way. The result of SPCA is a multi-band new spatial dataset with the same number of bands as the original data. The first principal component will have the greatest variance, the second will show the second most variance not described by the first, and so on. SPCA has certain advantages over conventional orthogonal functions, since they are not of any predetermined form, but are developed as unique functions from the data matrix [19].

In this paper, SPCA is introduced into the land sustainable use evaluation, and the formula for defining the land sustainable use evaluation is defined as follows Eq. (2):

$$E = \sum_{i=1}^{m} \sum_{j=1}^{n} (a_{ij}F_j) \tag{2}$$

where E represents the sustainable land use level; a_{ij} is the jth principal component corresponding to the ith raster; F_j is the eigenvalue contribution rate of the jth principal component.

3   Results

3.1   Spatial Distribution of the Indicators for Sustainable Land Use Evaluation

The spatial distributions of 13 indicators based on 5 km^2 are shown in Fig. 1. In the economy criteria layer (a–d), the trends of sustainable use of GDP and investment in fixed assets are similar, showing that areas with higher levels of sustainable land use

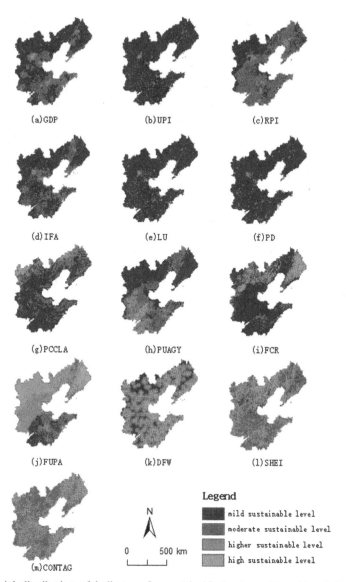

Fig. 1. Spatial distribution of indicators for sustainable land use (the abbreviations for the indicators are shown in Table 1)

are concentrated in coastal areas, Handan and Shijiazhuang; the sustainable land use level of rural income per capita indictor presents a decreasing trend from the southeast coast to the northwest inland; the sustainable land use level of urban income per capita indictor presents an overall low trend. In the social criterion layer (e–h), the sustainable land use level trend of land urbanization rate and population density show that the municipalities and provincial capitals have higher levels of sustainable land use. The level of sustainable land use in per unit area grain yield generally shows a downward trend from southwest to northeast; the sustainable land use level in per capita cultivated area showed a decreasing trend from northwest to southeast. In the ecological criterion layer (i–k), areas with high levels of sustainable land use indicated by forest coverage are mainly concentrate in the north with the highest level of sustainable land use in the northeast; areas with high level of sustainable use of fertilizers per unit of arable land area are concentrated in the northeastern and northwestern regions, and the sustainable land use level in Shandong is low. As for the indictor of distance to waterbodies, areas with higher levels of sustainable land use are all around waterbodies. The farther away from waterbodies, the lower the level of sustainable land use. In the ecological pattern criterion layer (l-m), the sustainable land use levels of the SHEI and CONTAG show spatial heterogeneity.

3.2 Spatial Evaluation of Sustainable Land Use

The cumulative contribution rate of the first five principal components reached 82.8%, that is, the information contained in these five components accounted for 82.8% of the total information contained (Table 2). Thus, the first five principal components can effectively summarize the comprehensive information on the level of sustainable land use. The corresponding original evaluation factor loads are shown in the Table 3.

Table 2. Eigen values and accumulative contribution rates of principal components

PC layer	Eigen value	Percent of eigen values	Accumulative of eigen values
1	1.1489	32.4262	32.4262
2	0.59281	16.7314	49.1576
3	0.44832	12.6534	61.8110
4	0.43119	12.1697	73.9806
5	0.31147	8.7910	82.7716
6	0.16481	4.6516	87.4232
7	0.12985	3.6648	91.088
8	0.09211	2.5997	93.6877
9	0.07857	2.2176	95.9054
10	0.05274	1.4885	97.3938
11	0.0472	1.3321	98.7259
12	0.02685	0.7578	99.4837
13	0.01829	0.5163	100

Table 3. Load matrix of principal components

Criteria	Indicators	PC1	PC2	PC3	PC4	PC5
Economy	IFA	−0.13483	0.32195	0.14059	−0.31068	0.22656
	GDP	−0.16734	0.23899	0.19029	−0.3169	0.10227
	RPI	−0.26857	0.03696	0.01579	−0.11095	−0.16213
	UPR	−0.02919	0.19305	0.09455	−0.20993	0.19538
Society	LU	−0.03597	0.24053	0.13493	−0.22974	0.26665
	PD	0.02322	−0.10513	−0.05936	0.11797	−0.09739
	PCCLA	−0.4351	−0.33687	−0.00659	−0.01061	−0.05345
	PUAGY	0.11146	−0.00267	−0.44011	0.43671	0.48363
Ecology	FCR	0.54574	−0.19202	0.40108	−0.0728	−0.39516
	FUPA	0.57818	0.29609	−0.01549	−0.01649	0.22384
	DFW	−0.21036	0.54738	0.38542	0.666	−0.22211
pattern	SHEI	0.01872	0.36995	−0.50781	−0.16258	−0.43441
	CONTAG	−0.02342	−0.23912	0.38967	0.11655	0.33483

Note: the abbreviations for the indicators are shown in Table 1

Based on the first five principal components, the spatial distribution characteristics of land sustainable use level in the study area are shown in Fig. 2. High level sustainable land use areas are mainly concentrated in the northeast, northwest and coastal area of the Bohai Rim. The land types in the northeast and northwest of the region are mainly forest land and grassland, which are less disturbed by human activities. The land types in coastal areas are mostly urban and rural residential land, and the level of economic development is relatively high, so land use sustainability is relatively high. The areas with low sustainable land use levels are mainly concentrated in the southwest. The land types in this area are mostly cultivated land, and the level of land development and utilization is weak.

The area of completely sustainable land use area is 109375 km^2, accounting for 21.39%, mainly distributed in the northern of Hebei Province, and the eastern of Liaoning and Beijing. The area of the land with higher sustainable utilization level is 145325 km^2, accounting for 28.42%, mainly distributed in Tianjin, Zhangjiakou, Chaoyang, Qinhuangdao, Anshan, etc. The area of the basic sustainable land use level is 166600 km^2, accounting for 32.58%, which is mainly distributed in the south of Hebei and the west of Liaoning Province. The area of initial sustainable land use level is 89750 km^2, accounting for 17.55%. It is mainly distributed in Linyi, Rizhao, Qingdao, Yantai, Weihai, Dongying, Binzhou, Laiwu and Zibo in the central and eastern parts of Shandong Province (Table 4).

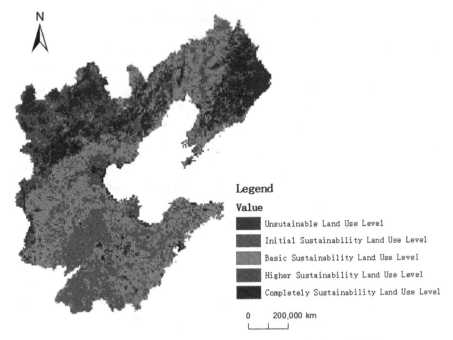

Fig. 2. Spatial pattern of sustainable land use level in the Bohai Rim

Table 4. Sustainable land use level results evaluation

Sustainable land use level	Level	Area (km^2)	Proportion (%)
Unsustainable level	1	275	0.05
Initial sustainability level	2	89,750	17.55
Basic sustainability level	3	166,600	32.58
Higher sustainability level	4	145,325	28.42
Completely sustainability level	5	109,375	21.39

4 Discussion

The level of sustainable land use in the region was influenced by many factors. The results of the study on the ecological criterion level indicate that the area of forest coverage and the amount of fertilization are the dominant determinants, which are related to the water and soil conservation capacity of the forest land and the influence of chemical fertilizers on the fertility of the soil. In the social criterion layer, the per capita arable land area is the dominant factor affecting the level of sustainable land use, which is closely related to the total population in the region, the available arable land area, and the stability of arable land. The higher the food production, the higher the level of sustainable land use. Locally, the level of sustainable land use in urban areas is generally higher than that in rural areas. The reason is that investment in fixed assets and GDP has increased the level of sustainable land use in urban and rural areas mainly from the socio-economic aspects.

In order to further improve the level of sustainable land use in the Bohai Rim region, the government should take corresponding effective measures against weak links. At the ecology level, the resource types with higher benefits such as forest land and waters are defined as high-quality land resources. In order to maintain their long-term availability and stability, the government should make overall plans for land use and rationally optimize the development and utilization structure of land resources. Comprehensively considering economic, social, and other factors, give full play to the utilization efficiency of high-quality land resources and increase the economic benefits of land and nature. In view of the low level of sustainable land use in the south, we should vigorously promote land development and land reclamation, increase the effective arable land area in Shandong, consider spatial differences in agricultural productivity, and highlight regional land use characteristics to improve sustainable land use level. The increasing population has caused the demand for cultivated land to increase. The analysis of the demographic composition in the social norms indicates that the sustainable land use level of municipalities and provincial capitals is low. In response to the problem of sustainable land use in urban areas, the government should reasonably control the population, increase the carrying capacity of arable land through technical means, and improve the overall level of sustainable land use.

5 Conclusion

This paper established a four-dimensional evaluation model of sustainable land use, namely "Economy-Society-Ecology-Pattern" framework, and proposed a novel method to evaluate the land use sustainability at an explicit spatial level. First, based on the "Economy-Society-Ecology-Pattern" framework, an index system of sustainable land use evaluation with 13 indicators was established. Then, the indicators were discretized at 5 km grids level with spatial analysis and land use data. At last, the spatial principal component analysis method is used to get the comprehensive sustainability level of the region. The level of sustainable land use in the Bohai Rim is generally at the stage of basic sustainable development. High level sustainable land use areas are mainly concentrated in the northeast in Liaoning and northwest in Hebei. The ecology factor has the strongest impact on the sustainable land use level, and the economy factor has the weakest impact on the sustainable land use level.

The method could offer more spatial information of sustainable land use to help government propose more efficient policies regarding improving land use sustainability. Due to the limitation of the data acquisition, the sustainable land use index system constructed in this paper need to get improved in the future. If the soil quality index could be added into the index system, the reliability of the research results would be more convincing.

References

1. Zhang, Z.F.: Sustainable land use goals, challenges and strategies for SDGs. Land Sci. China **33**(10), 48–55 (2019). https://doi.org/10.11994/zgtdkx.20191010.140058

2. Smyth, A.J., Dumanski, J.: FESLM: An international framework for evaluating sustainable land management. Land and Water Development Division, FAO, Rome (1993). https://doi. org/10.4141/cjss95-059

3. Hurni, H.: Assessing sustainable land management (SLM). Agr. Ecosyst. Environ. **81**(2), 83–92 (2000). https://doi.org/10.1016/S0167-8809(00)00182-1

4. Fu, B.J., Chen, L., Ma, C.: Index system and method for sustainable land use evaluation. J. Nat. Resour. **12**(2), 112–118 (1997) https://doi.org/10.11849/zrzyxb.1997.02.003

5. Peng, B.Z., et al.: Research on sustainable use of land resources in the Yangtze river delta. J. Nat. Resour. **16**(4), 305–312 (2001). https://doi.org/10.3321/j.issn:1000-3037.2001.04.002

6. Zhou, B.Z., et al.: PSR model and its application insustainable land use evaluation. J. Nat. Resour. **17**(5), 541–548 (2002). https://doi.org/10.3321/j.issn:1000-3037.2002.05.003

7. Xie, H.L., et al.: Regional land use sustainability level measurement based on PSR model—taking Poyang lake ecological economic zone as an example. Resour. Sci. **37**(3), 449–457 (2015)

8. Chen, B.: Construction and evaluation of the regional land sustainable use index system framework. Adv. Geogr. Sci. **21**(3), 204–215 (2002). https://doi.org/10.3969/j.issn.1007-6301.2002.03.002

9. Chen, B.M., Zhang, F.R.: Theories and methods of sustainable land use index system in China. J. Nat. Resour. **16**(03), 197–203 (2001). https://doi.org/10.3321/j.issn:1000-3037. 2001.03.001

10. He, C., et al.: Evaluation of sustainable land management in urban area: a case study of Shanghai China. Ecol. Indic. **80**, 106–113 (2017). https://doi.org/10.1016/j.ecolind.2017. 05.008

11. Zhang, X., Wu, Y., Shen, L.: An evaluation framework for the sustainability of urban land use: a study of capital cities and municipalities in China. Habitat Int. **35**(1), 141–149 (2011). https://doi.org/10.1016/j.habitatint.2010.06.006

12. Banai, R.: Land resource sustainability for urban development: spatial decision support system prototype. Environ. Manage. **36**(2), 282–296 (2005). https://doi.org/10.1007/s00267-004-1047-0

13. Peng, J., et al.: Research progress on land use evaluation at home and abroad. Resour. Sci. **25**(02), 87–95 (2003). https://doi.org/10.3321/j.issn:1007-7588.2003.02.014

14. Liu, W.B., Zeng, J.X.: Changes in my country's population and urbanization spatial distribution since 2000. J. South China Normal Univ. (Nat. Sci. Edn.) **47**(04), 108–115 (2015). https://doi.org/10.6054/j.jscnun.2014.12.026

15. Yang, R., Liu, Y.S., Long, H.L.: Co-evolution characteristics of population-land-industry non-agricultural transformation in Bohai Rim region of China. Geogr. Res. **34**(03), 475–486 (2015). https://doi.org/10.11821/dlyj201503007

16. Li, Y.H., Chen, C., Liu, Y.S.: Measurement and types of urban-rural development transformation in China—take Bohai Rim Region as an example. Geogr. Res. **33**(09), 1595–1602 (2014). https://doi.org/10.11821/dlyj201409001

17. Yang, Y., et al.: Comparison of spatial and temporal dynamics of urban land use and population size distribution in China—take Bohai Rim as an example. Geogr. Res. **35**(9), 1672–1686 (2016). https://doi.org/10.11821/dlyj201609007

18. Xu, X.: Dataset of the spatial distribution of GDP in China at 1 km^2 grid. Data registration and publishing system of data center of resources and environment science, https://doi.org/ 10.12078/2017121102

19. Rejaur, R., Shi, Z.H., Cai, C., Zhu, D.: Assessing soil erosion hazard -a raster based GIS approach with spatial principal component analysis (SPCA). Earth Sci. Informat. **8**(4) (2015). https://doi.org/10.1007/s12145-015-0219-1.

ST-GWANN: A Novel Spatial-Temporal Graph Wavelet Attention Neural Network for Traffic Prediction

Zunhao Liu[1], Zhiming Ding[2(✉)], Bowen Yang[1], Lei Yuan[1], Lutong Li[1], and Nannan Jia[1]

[1] Beijing University of Technology, Beijing 100124, China
{bovin.y,yuanlei,lilutong,jianannan}@emails.bjut.edu.cn
[2] Institute of Software, Chinese Academy of Sciences, Beijing 100190, China
zhiming@iscas.ac.cn

Abstract. In a modern Intelligent Transportation System (ITS), traffic prediction exerts an enormous function in alleviating traffic congestion and path planning. Due to the complex dynamic spatial-temporal dependence of traffic data, the traditional prediction methods have some limitations in the spatial-temporal correlation modeling, and cannot effectively predict both long-short term traffic conditions. In this paper, a novel deep learning framework Spatial-Temporal Graph Wavelet Attention Neural Network (ST-GWANN) is proposed for long-short term traffic prediction, which can comprehensively capture the spatial-temporal features. In the framework, the graph wavelet neural network and attention mechanism are integrated into the spatial gated block, which can obtain the spatial dependence of the road network. By combining the Gated Linear Units (GLU) and the temporal transformer layer, the local and global dependence of the time dimension are obtained. The proposed framework experimented on two real-world datasets, the results show that ST-GWANN outperforms state-of-art methods in traffic prediction tasks.

Keywords: Traffic prediction · Graph wavelet · Spatial-temporal model · Transformer

1 Introduction

A significance in urban development has been played by the Intelligent Transportation Infrastructure (ITS). Traffic congestion in many cities has been increasingly serious with the pace of urbanization, which has severely decreased the work and travel efficiency of people. Accurate and timely multi-scale traffic prediction can effectively solve the urban traffic congestion issue, optimize people's path planning, while saving people's travel time [1], and can provide urban management agencies with great support to solve traffic planning problems [2].

© Springer Nature Switzerland AG 2021
G. Pan et al. (Eds.): SpatialDI 2021, LNCS 12753, pp. 83–99, 2021.
https://doi.org/10.1007/978-3-030-85462-1_7

Traffic prediction is to predict the future traffic speed given the historical traffic speed and the basic road network. Traffic prediction is split into two scales according to the prediction time: short-term (0–30 min) and medium-term (more than 30 min) [3]. Improving the accuracy of traffic prediction is also a subject that many researchers are devoted to studying, which has tremendous importance for application. However, because of its dynamic spatial-temporal correlation, traffic prediction poses the following challenges:

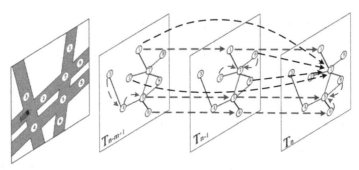

Fig. 1. The picture (left) shows a road network with 9 sensors. The picture (right) shows the road traffic from time T_{n-m+1} to time T_n.

1   The time dependence of the traffic road network is very strong, and the traffic condition varies regularly and trendily over a period of time. During the morning and evening peak hours during working days and after emergencies, the traffic condition on the road varies dynamically. This brings challenges to long-term predictions. As shown by sensors 1, 5, and 6 in Fig. 1, the future traffic condition of the same sensor is linked to the traffic condition at historical moments. At the same time, the traffic situation of the same sensor at different times in history will also be related to the traffic situation of its associated node at different times in the future. As shown in Fig. 1, sensors 1 and 2 at time T_{n-m+1} and sensors 8, 9 at time T_{n-1} have an impact on sensor 7 at time T_n in the future.

2   The spatial dependency of the road network on traffic is intense. Roads of different levels have different carrying capacities in the actual road network, and the traffic conditions of two roads that are close in spatial structure may be completely different. Moreover, the upstream road traffic and the downstream road traffic feedback to each other. This means that the traffic network's topological structure is non-Euclidean, and how to obtain its spatial dependency efficiently is also a challenge. As shown by the blue dotted line in Fig. 1, at time T_n, the road vehicles of sensors 6, 8 will merge into the road where the adjacent sensor 5 is located, making the road where sensor 5 is located becomes congested.

Many sensors are deployed on the road to gather traffic information as the Internet of Things technology evolves. These sensors can capture and produce time series data continuously, which lays the groundwork for traffic prediction. In recent years, researchers have been proposing more traffic prediction methods based on deep learning, such as Long Short-Term Memory (LSTM) [4] and Convolution Neural Network (CNN) [5], but these methods cannot obtain the spatial-temporal dependence of traffic conditions at the same time. In particular, the rise of graph neural networks also offers a new way to model the spatial dependency of road networks. For example, STGCN [3] uses spatial-temporal convolution block iteration to forecast, which tends to accumulate errors in long-term predictions. DCRNN [6] makes predictions by combining graph neural networks and recurrent neural networks, but the results obtained through only one evaluation cannot be a good balance between long and short-term prediction tasks.

In view of the limitations of existing work, we propose a new Spatial-Temporal Graph Wavelet Attention Neural Network (ST-GWANN) to solve the task of traffic prediction. First, the Graph Convolutional Network (GCN) and the attention mechanism are combined into a spatial gated block through a gating method. Through the gated block, the spatial adjacency relationship of the road network can be better obtained. In addition, in order to perform convolution more effectively For calculation, we use the Graph Wavelet Transform [7] to replace the Laplace transform in the original graph convolution, so that the graph convolution avoids the eigenvalue decomposition and makes it have higher computational efficiency. Inspired by transformer [8], transformers can efficiently model long-term dependence and effectively reducing the influence of errors caused by multiple iterations. At the same time, by combining Gated Linear Units (GLU) [25], the model can accurately obtain the local and long-term characteristics of the time dimension. Experiments on real-world datasets show that our model is superior to existing baseline methods. The main contributions of this paper are summarized as follows:

1. A novel deep learning framework ST-GWANN is proposed in this paper, which combines GLU and the temporal transformer layer to capture the local and global temporal dependence, facilitating the prediction of long-short term dependence.
2. A graph wavelet neural network is used to capture the topology of a road network, incorporate an attention mechanism, and use a gating mechanism to combine them to model the spatial dependence of the road network.
3. The proposed model is evaluated on two real-world traffic datasets and compares with the existing baseline model, it is proved that the model is superior in long- short term traffic prediction tasks, and has and has good robustness in long term prediction.

The remainder of this paper is structured as follows. Related work on traffic prediction is reviewed in Sect. 2. Section 3 introduce some preliminary traffic prediction concepts. Section 4 outlines the specifics of our proposed framework for traffic prediction. The experimental methods and results are introduced in Sect. 5. This paper is summarized in Sect. 6.

2 Related Work

Traffic prediction methods are mainly divided into knowledge-driven methods and data-driven methods. Knowledge-driven methods usually apply to queue theory to simulate traffic conditions through users' prior knowledge. The early data-driven methods mainly include Support Vector Regression (SVR) [9], but this method is limited to stationary time series [22]. The Auto-Regressive Integrated Moving Average (ARIMA) [10] model is extended to deal with complex time series. But long-term time-series data can still not be handled well. In recent years, the development of deep learning methods has made data-driven methods show good performance. Recurrent Neural Networks (RNN) [11] and its variants Long Short-term Memory Network (LSTM) [12] and Gated Recurrent Unit (GRU) [13] have been very helpful in dealing with the non-correlation of time series data [23], but it is susceptible to the inherent sequence of long-term series. GLU [25] solves efficiency issues but performs poorly in terms of scalability. The transformer [14] with a self-attention mechanism directly learns the dependence between each pair of input and output positions through the multi-head attention mechanism, which is better than the recurrent neural network. But the above models do not consider the spatial structure of the road network.

Zhang et al. [15] combined CNN and LSTM to model two-dimensional spatial-temporal traffic data to obtain spatial-temporal correlation, but CNN cannot handle non-European structure road network data well. Therefore, GCN [16, 17] is used to model non-Euclidean spatial structure data, which is more in line with the actual traffic network structure. Seo et al. [18] proposed Graph Convolutional Recurrent Network (GCRN) that combines GCN and RNN, a deep learning model that can predict structured sequence data, and combines structured sequence modeling. Yu et al. [3] proposed a Spatial-Temporal Graph Convolutional Network (STGCN) solve the problem of spatial-temporal data prediction in the transportation field, but the prediction performance of single-step prediction will be significantly reduced. Wu et al. [19] proposed Graph Wavenet to accurately capture the spatial correlation hidden in the data. The application of the attention mechanism [24] can more effectively deal with the temporal and spatial dynamic correlation of traffic data [20], but it is still difficult to determine the appropriate neighborhood of the central vertex. Xu et al. [7] proposed a Graph Wavelet Neural Network, which uses a thermal diffusion wavelet to obtain the relevant neighbors of the core node, which improves the efficiency of graph convolution, but it cannot process time-series data. In order to overcome these problems, a spatial gated block based on graph wavelet convolution and attention mechanism is established. Then, a more effective long-short term time series prediction is made by combining GLU and the temporal transformer.

3 Preliminaries

This section defines the traffic prediction problem and the traffic network.

3.1 Traffic Prediction Problem

Problem Definition. The traffic prediction problem is based on historical traffic observations (such as traffic flow, speed, etc.) detected by N traffic sensors on the road at the same frequency, which can be expressed as a time series of M historical moments on the traffic network. The multi-step prediction of the traffic situation in the future T time can be defined as.

$$\hat{x}_{t+T}, \hat{x}_{t+1} = \underset{x_{t+1},\dots,x_{t+T}}{argmax} \ logP(x_{t+T},\dots,x_{t+1}|x_t,\dots,x_{t-M+1}) \tag{1}$$

Where $x_t^i \in R$ represents the historical traffic observation value of the i-th node (sensor) at time step t, $X = \left[x_t,\dots,x_{t-M+1}\right]^T \in R^{N \times \tau}$ represents all historical traffic observations of N nodes (sensors) at time step t. Traffic prediction can be aimed at traffic speed, traffic flow, etc. In this paper, we use traffic speed as a feature of traffic conditions.

Traffic Network. In this paper, the traffic network is represented as a graph $G = (V, E, A)$, V represents the set of nodes (sensors) in the road network and $|V| = N$, corresponding to the traffic recorded by the actual road network nodes Observations; E represents the edge set of connectivity between sensors; A is the weight matrix, and $A \in R^{N \times N}$ represents the adjacency matrix constructed by the Euclidean distance between nodes. The Laplacian matrix L of graph G is defined as $L = D - A$, and D is the degree matrix of graph G, specifically $D_{ii} = \sum_j A_{ij}$. The normalized Laplace matrix is $L = I_n - D^{-1/2}AD^{-1/2}$, where I_n is the Identity Matrix.

4 The Proposed Framework

The proposed ST-GWANN framework is described in this section. Specifically, a network architecture is described first, followed by a spatial gated block, GLU, and temporary transformer.

4.1 Network Architecture

An iterative prediction mechanism is used by the ST-GWANN framework, which consists of four layers as shown in Fig. 2, namely the spatial gated block, the GLU layer, the transformer layer, and the prediction layer. A graph convolution module and an attention module are contained in the spatial graph convolutional neural network layer. Inspired by the gating mechanism [13], we use a gate structure to integrate the two modules. The attention module is used to extract road correlation. The GLU layer is used to capture local time dependence, and each GLU is composed of one-dimensional convolution. The transformer layer includes a multi-head attention layer, a forward feedback layer, a residual, and a normalization layer, and uses the value from the GLU as input to capture global dependence. Finally, the output of the transformer layer is processed by the prediction layer to predict and output the spatial-temporal series data. Next, we first discuss modeling spatial correlation with spatial gated block and then describe how to use GLU layer and transformer layer to capture temporal correlation.

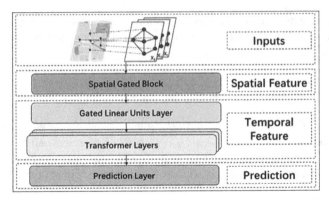

Fig. 2. The proposed spatial-temporal graph wavelet attention neural network framework.

4.2 Spatial Gated Block for Modeling the Spatial Features

We extract the relationships between roads through the convolution module, with some details as follows. Graph convolution neural networks are mainly divided into spectral and spatial methods. The spectral method of graph convolution is to convolute a graph signal using a diagonal linear operator defined in the Fourier domain. According to the definition of the convolution theorem, a graph convolution operation on $x \in R^N$ can be represented as:

$$x *_{\mathcal{G}} y = g_\theta(L)x = g_\theta\left(U \Lambda U^T\right)x = U g_\theta(\Lambda) U^T x \tag{2}$$

$*_{\mathcal{G}}$ is a graph convolution operator defined in the Fourier domain, y represents a filtering kernel, $U = (u_1, u_2, \ldots, u_n) \in R^{N \times N}$ is a Fourier basis composed of feature vectors, $g_\theta(\Lambda)$ is a diagonal matrix composed of the eigenvalues of the Laplace matrix $g_\theta(\Lambda) = \sum_{k=0}^{K} \theta_k \Lambda^k$. However, when the scale of the graph to be convolved becomes larger, the eigendecomposition of the Laplacian matrix in Eq. (2) requires a computational cost of $O(n^3)$, which makes the Fourier transform of the graph inefficient. [21] Use Chebyshev polynomial expansion to solve the above problems:

$$g_\theta(\Lambda) = \sum_{k=0}^{K-1} \theta_k T_k\left(\tilde{L}\right) \tag{3}$$

Where θ_k is the Chebyshev coefficient vector, $\tilde{L} = \frac{2}{\lambda_{max}}L - I_N$, λ_{max} represents the maximum value of the Laplacian matrix, I_N is the identity matrix, $T_k(x) = 2xT_{k-1}(x) - T_{k-2}(x), T_0(x) = 1, T_1(x) = 1$ represents the K-order Chebyshev polynomial expansion, which is graphically equivalent to extracting the K-1 order neighbor node information for each node to compute node characteristics. However, the size of the K-order limits the flexibility of graph convolution. In this paper, graph wavelet transform is used instead of graph Fourier transform to extract spatial correlation more flexibly. Wavelet transform defines $\psi_s = \{\psi_{s1}, \psi_{s2}, \ldots, \psi_{sn}\}$ is the base, ψ_{si} is the energy diffused from the i-th node, and s is the diffusion range. The larger the s, the larger the diffusion range. The eigenvector of the wavelet base ψ_s dependent Laplacian matrix can be expressed as:

$$\psi_s = UG_sU^{\mathrm{T}} \tag{4}$$

Where $G_s = \mathrm{diag}\big(\{g_s(\lambda_i)\}_{i=1}^n\big) \in R^{N \times N}$ is a scaling matrix, which is applied to the eigenvalue λ_i through the g function to obtain diagonal elements, in In this paper, we use the thermal kernel function $g_s(\lambda_i) = e^{s\lambda_i}$. Similarly, the matrix of the graphic wavelet inverse transformation can be expressed as $\psi_s^{-1} = UG_{-s}U^{\mathrm{T}}$, the wavelet on the graph The transformation is expressed as $\hat{x} = \psi_s^{-1}x$, and the graph wavelet transform is used to replace the graph Fourier transform in Eq. (2), we get the graph convolution as follows:

$$x *_G y = \psi_s\Big(\big(\psi_s^{-1}y\big) \odot \big(\psi_s^{-1}x\big)\Big) = \psi_s g_\theta(G_s)\psi_s^{-1}x \tag{5}$$

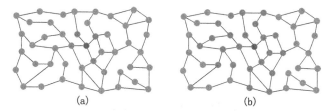

Fig. 3. Examples of wavelets at different scales, (a) for small scales and (b) for large scales.

Figure 3 shows an example of a wavelet on different scales, showing two different scales of wavelet bases. The red node represents the center node, the green node represents the diffusion range of the wavelet bases at different scales, the blue node represents the other connected nodes, and each edge represents the connected road between the two nodes. Compared with the small-scale wavelet base in Fig. 3(a), the large-scale wavelet base in Fig. 3(b) have larger perception fields. We derive the spatial characteristics using the topological relationships in which the signals diffuse from the central node (Eq. (5)).

Compared with spectral convolution neural network, wavelet transform has the following advantages: (1) pass through the wavelet base ψ_s and ψ_s^{-1} replaces U and U^{T} of Fourier base because the wavelet transform avoids the characteristic decomposition of Laplace matrix; (2) Localized convolution and high spareness of the wavelet base greatly reduces the computational complexity of the graphical convolution neural network and makes the computational process more efficient; (3) the node neighborhood can be flexibly adjusted by the change of scale s.

In the above calculation, the road information is obtained by the geographic proximity of the nodes, but in the real road case, the influence of neighboring nodes around each node is different. In this paper, we introduce the attention module to aggregate the characteristics of neighbor nodes to capture the spatial correlation of nodes more effectively. The input of the attention module is the eigenvector $q = \{\vec{q_1}, \vec{q_2}, \cdots, \vec{q_N}\}$, $\vec{q_i} \in R^P$, where the node represents the sensor, q represents the set of eigenvectors of the sensor, and p represents the number of eigenvectors of each node. Next, a self-attention mechanism is implemented for each node, and the attention coefficients are calculated as shown in Eq. (6):

$$e_{ij} = a(\overrightarrow{q_i} w, \overrightarrow{q_j} w) \tag{6}$$

$a(\cdot)$ is a function of the attention mechanism, $w \in P \times P'$ is a weight matrix, and α_{ij} in Eq. (6) indicates how much node j affects node i. We normalize the attention factor of the node neighborhood using a non-linear function and get the output result A_i as shown in Eq. (7) and Eq. (8):

$$\alpha_{ij} = sigmoid(e_{ij}) \tag{7}$$

$$A_i = \sum_{j \in \tilde{N}_i} a_{ij} \overrightarrow{q_j} w \tag{8}$$

Inspired by the gate structure of LSTM [12] and GRU [13], the graph convolution module and the attention module are integrated using a gate structure. First, pass the input to the graph convolution module and the attention module, and then use the sigmoid function to map the result of the graph convolution module to a value between 0 and 1, and the output is the element-wise Hadama product of the convolution and attention module results.

4.3 Gated Linear Units Convolution Layer for Extracting Local Temporal Features

After modeling the spatial correlation with the spatial graph convolutional neural network layer, we use GLU to capture the local temporal feature dependence. GLU is a gating mechanism in convolutional neural networks. The difference from Gated Recurrent Unit (GRU) [13] processing time-series information is that GLU gradient propagation is easier and the gradient disappearance phenomenon is improved. And the computational efficiency has also been greatly improved.

GLU contains 1-D convolution with a width-Kt kernel, gated linear unit, and residual connections, input $X = \{x_1, x_2, \ldots, x_i, \ldots, x_N\}$, $x_i \in R^{N \times T \times C_i}$, get the convolution result $Y = [AB] \in R^{N \times (T-K_t+1) \times 2C_o}$ through the convolution kernel $\theta \in R^{K_t \times C_i \times 2C_o}$, C_o is the size of the feature set generated by GLU. The output of the 1D convolutional layer is divided into two parts of the same size. The equation through the gating mechanism is as shown in Eq. (9):

$$H = A \odot \sigma(B) \in R^{N \times (T-K_t+1) \times C_o} \tag{9}$$

A and B are the input parts of GLU, σ represents the sigmoid function, and $\odot$ represents the element-wise Hadamard product of A and B. The output of Eq. (8) is input to the residual connection in order to prevent the gradient from disappearing, if the number of input dimensions does not match the number of output dimensions, input zeros are used as padding.

4.4 Temporal Transformer for Extracting Global Temporal Features

We can obtain the local temporal features by calculating the convolution layer of GLU. However, in the actual traffic condition, the time feature may be very complex, and it is not complete to obtain the feature only by GLU. Therefore, we can obtain the global time feature dependency by inputting the temporary transformer [8].

The transformer is a kind of neural network model for long sequence modeling. Different from the RNN based modeling models such as LSTM [12], transformer completely adopts the attention mechanism to realize sequence modeling, which improves the shortcomings of RNN such as gradient explosion and error accumulation in long sequence modeling. Parallel computing also improves its operational efficiency.

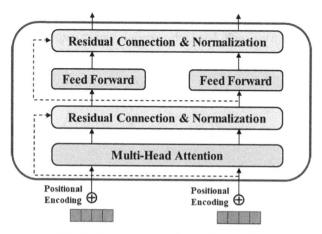

Fig. 4. Temporary transformer framework.

As shown in Fig. 4, the temporary transformer mainly includes position coding, multi-head attention layer, residual connection & Layer Normalization layer, and feed-forward output layer. The transformer layer input is a matrix generated by stacking each GLU unit with the output of the sequence. It is then passed to the attention layer after the input is position-encoded. The attention layer's function is to map output to a query and to a set of key-value pairs. Specifically, the input of the attention layer consists of the query matrix $Q \in R^{T \times d_k}$ of all positions in the sequence, and the dimension is d_k The key matrix $K \in R^{T \times d_k}$ and the value matrix $V \in R^{T \times d_v}$ with dimension d_v are composed of a scaled dot-product attention mechanism for output calculation, as shown in Eq. (10):

$$Attention(Q, K, V) = softmax\left(\frac{QK^T}{\sqrt{d_k}}\right)V \qquad (10)$$

Through the dot product of query matrix and key matrix in Eq. (10) and divide by $\sqrt{d_k}$, each location's attention score is normalized by *softmax* function, and the attention score is multiplied by the corresponding value matrix V as a weight to get the output.

The query matrix Q^T, the key matrix K^T, and the value matrix V^T need to be mapped for each spatial node $X^T \in R^{T \times d}$, as shown in Eq. (11):

$$Q^T = H^T W^Q, K^T = H^T W^K, V^T = H^T W^V \tag{11}$$

$W^Q \in R^{d \times d_k}$, $W^K \in R^{d \times d_k}$ and $W^V \in R^{d \times d_v}$ is the parameter matrix of the projection, and the result is taken into the Eq. (10) to obtain the output of the attention mechanism of this paper:

$$Attention\left(H^T\right) = softmax\left(\frac{\left(H^T W^Q\right)\left(H^T W^K\right)^T}{\sqrt{d_k}} H^T W^V\right) \tag{12}$$

Compared with a single attention function, multi-head attention can simultaneously focus on spatial feature information from different locations. In order to improve the stability of training, we use multi-head self-attention, and its output is the connection of each individual attention function, as shown in Eq. (13):

$$
\begin{aligned}
Multihead\left(H^T\right) &= Concat\left(head_1, \ldots, head_j\right) W^O \\
whereheadj &= attention_j\left(H^T\right) \\
&= softmax\left(\frac{\left(H^T W_j^Q\right)\left(H^T W_j^K\right)^T}{\sqrt{d_k}} H^T W_j^V\right)
\end{aligned}
\tag{13}
$$

Where $W^O \in R^{hd_v \times d}$ is a linear output projection matrix, $W_j^Q \in R^{d \times d_k}$, $W_j^K \in R^{d \times d_k}$, $W_j^V \in R^{d \times d_v}$ are the projection matrices for the j-th attention head. After the output of the multi-head attention layer passes through a residual connection and normalization layer, it will be transferred to the feed-forward neural network layer, and then after a residual connection and normalization layer, the output $H_{OUT}^T \in R^{T \times d}$ of the transformer layer is obtained.

Finally, a prediction Layer is connected to the end of the transformer layer, and the output $H_{OUT}^T \in R^{T \times d}$ of the transformer layer is passed to the prediction layer composed of the feed-forward neural network, and the traffic information of the future T' time steps is output. We use Mean Square Error Loss (L2) as the loss function for overall model training, which is defined as shown in Eq. (14):

$$L\left(\hat{Y}; W_\theta\right) = \sum_t \hat{Y}(X_{t-T}, \ldots, X_t, W_\theta) - X_{t+1}^2 \tag{14}$$

Where $\hat{Y}(\cdot)$ represents the predicted traffic information, X_{t+1} represents the input traffic information, and W_θ represents all the parameters to be trained in this paper.

5 Experiments

In this section, we verify the validity of the model in traffic prediction tasks through experiments. First, the dataset is described, then the experimental settings, the relevant evaluation metrics, and the baseline model of experimental comparison are introduced. Finally, the experimental results are shown and discussed.

5.1 Dataset Description

We used two real-world datasets BJFR4 and PeMSD4, which were collected by Amap Open Platform [23] and the California Department of Transportation [20].

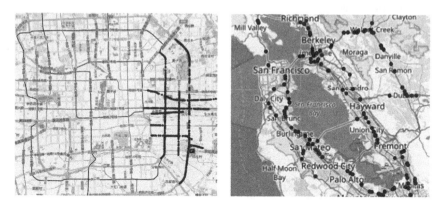

Fig. 5. Sensor distribution for BJFR4 (left) and PeMSD4 (right) datasets.

BJFR4. It collects traffic conditions (speed, congestion) on nine major roads near Central Business District in Beijing from September to October 2020. It contains 927 sensors with a sampling interval of 5 min. The sensor distribution is shown in the left image of Fig. 5.

PeMSD4. It collects traffic data (traffic volume, speed, and lane share) for the San Francisco Bay area from January to February 2008. It contains 307 sensors with a sampling interval of 5 min, and the sensor distribution is shown in the right image of Fig. 5.

The dataset contains 288 pieces of data per day, and the missing values of traffic data are filled in by linear interpolation. We divide the dataset into training set, validation set, and test set at a scale of 6:2:2.

5.2 Experimental Settings

We will conduct five comparative experiments: (1) A comprehensive comparison between ST-GWANN and six baseline models. (2) Verify the influence of the temporary transformer layer on long-term traffic prediction. (3) Verify the influence of the graph wavelet neural network on the calculation efficiency. (4) Verify the extraction of the attention mechanism on the spatial dependence of the traffic road network. (5) Show the visual comparison between the ST-GWANN prediction results and the ground truth.

The experimental equipment is Dell Precision 7920 Workstation (CPU: Intel Xeon (R) Silver 4210 @2.20GHz, GPU: NVIDIA GeForce RTX 2080 Ti, RAM: 128.0GB). We set the initial learning rate to 0.001. Starting with the 20th epoch, the learning rate decays to 1/10 of the original every 10 epochs, Batch-Size is set to 64 in training, Adam is selected as the optimizer, the number of convolution cores is 64, we set the elements of

ψ_s and ψ_s^{-1} smaller than a threshold t = {0.0001, 0.0001} to 0, the number of transformer layers set to 1, attention head set to 4, training stops after verifying that the loss has not been reduced for 200 consecutive periods.

5.3 Evaluation Metrics and Baselines

Mean Absolute Error (MAE), Mean Absolute Percentage Error (MAPE), Root Mean Square Error (RMSE) are used as the evaluation metrics, as shown in Eq. (15), Eq. (16) and Eq. (17):

$$MAE = \frac{1}{n} \sum\nolimits_{t=1}^{n} |x_t - \tilde{x}_t| \tag{15}$$

$$MAPE = \frac{1}{n} \sum\nolimits_{t=1}^{n} \left| \frac{x_t - \tilde{x}_t}{x_t} \right| \times 100 \tag{16}$$

$$RMSE = \left[\frac{1}{n} \sum\nolimits_{t=1}^{n} (x_t - \tilde{x}_t)^2 \right]^{\frac{1}{2}} \tag{17}$$

Where $\tilde{x}_t$ represents the predicted value, x_t represents the true value, and n is the number of predicted values.

In order to verify the efficiency of ST-GWANN, ST-GWANN will be compared with some representative methods of traffic prediction. The baseline is as follows:

HA: Use historical averages as forecasts for future cycles.

SVR[9]: Use support vector machine to perform regression analysis and prediction.

ARIMA[10]: Auto-regressive integrated moving average is widely used in time series prediction.

FC-LSTM[12]: Fully connected long short term memory network.

STGCN[3]: A spatial -temporal graph convolution model.

ASTGCN[20]: An attention-based spatial-temporal graph convolutional network for spatial-temporal sequence prediction.

5.4 Experiments

The comparison of the traffic prediction results of ST-GWANN and six baseline models in the BJFR4 dataset and PeMSD4 dataset is shown in Table 1 and Table 2. We averaged the 15-min, 30-min, and 60-min prediction results and highlighted the best results. From the table, we can see that ST-GWANN performed significantly better on two real-world datasets than the baseline method, and the values of MAE, MAPE, and RMSE were the lowest, especially on the long-short term predictions of PeMSD4. Compared with the best baseline model, the evaluation improved by 11.4%, 7.69%, and 5.05%, respectively.

In Table 1 and Table 2, the traditional prediction methods (i.e., HA, ARIMA and SVR) have limited ability to process complex traffic data, leading to unsatisfactory prediction results. Compared with traditional prediction methods, the traditional deep learning method LSTM can handle complex traffic data, but it performs poorly due to the accumulation of errors in long-term prediction. The method based on graph neural networks can effectively handle spatial-temporal correlation. STGCN, ASTGCN, and ST-GWANN have achieved remarkable results compared with traditional methods. ASTGCN and ST-GWANN use the attention mechanism to consider the neighbor information of the road network than STGCN, so they perform better on the two data sets, but ASTGCN has no advantage in the long-term prediction. Compared with other baseline models, ST-GWANN can effectively capture the temporal and spatial characteristics of prediction tasks and is more pronounced in long-term prediction.

Table 1. Comparison of traffic prediction results on the BJFR4 dataset

Model	15 min			30 min			60 min		
	MAE	MAPE	RMSE	MAE	MAPE	RMSE	MAE	MAPE	RMSE
HA	5.57	13.97%	7.96	5.57	13.97%	7.96	5.57	13.97%	7.96
ARIMA	5.35	10.67%	8.34	5.84	14.84%	10.75	6.80	18.36%	13.98
SVR	5.29	10.16%	8.53	5.54	12.46%	11.24	7.30	15.62%	14.53
FC-LSTM	4.45	10.72%	6.65	4.83	11.81%	8.07	5.92	14.05%	10.16
STGCN	3.84	8.86%	5.53	4.30	10.36%	6.89	5.07	12.42%	8.23
ASTGCN	3.57	8.35%	5.34	4.06	9.67%	6.52	4.83	11.93%	7.88
ST-GWANN	**3.39**	**7.97%**	**5.15**	**3.77**	**9.25%**	**6.19**	**4.25**	**11.14%**	**7.41**

Table 2. Comparison of traffic prediction results on the PeMSD4 dataset

Model	15 min			30 min			60 min		
	MAE	MAPE	RMSE	MAE	MAPE	RMSE	MAE	MAPE	RMSE
HA	2.72	6.35%	5.53	2.72	6.35%	5.53	2.72	6.35%	5.53
SVR	1.87	3.74%	3.65	2.51	5.63%	5.12	3.41	8.02%	7.15
ARIMA	1.72	3.28%	3.19	2.48	5.19%	4.83	3.34	8.15%	6.71
FC-LSTM	2.11	4.64%	4.26	2.24	5.03%	4.74	2.71	5.59%	5.13
STGCN	1.55	3.02%	3.37	2.02	3.59%	3.89	2.46	5.07%	4.75
ASTGCN	1.43	2.83%	2.94	1.81	3.31%	3.58	2.18	4.68%	4.51
ST-GWANN	**1.22**	**2.31%**	**2.41**	**1.53**	**3.04%**	**3.34**	**1.93**	**4.17%**	**4.24**

Effect of Temporal Transformer. To further research the function of the temporal transformer layer, we compared the influence of ST-GWANN (GLU) with the excluded temporal transformer layer and the complete ST-GWANN on the PeMSD4 dataset in Table 3. We can see that with the extension of the prediction period, the transformer model owned by ST-GWANN has significant advantages in long-term prediction due to its complex network structure and better learning ability. Although ST-GWANN (GLU) does not perform as well as ST-GWANN, it is still superior to most baselines, further demonstrating the importance of the proposed components.

Table 3. Comparison between ST-GWANN(GLU) and ST-GWANN on PeMSD4 dataset

Model	15 min			30 min			60 min		
	MAE	MAPE	RMSE	MAE	MAPE	RMSE	MAE	MAPE	RMSE
ST-GWANN (GLU)	1.30	2.42%	2.56	1.61	3.29%	3.56	2.07	4.53%	4.62
ST-GWANN	**1.22**	**2.31%**	**2.41**	**1.53**	**3.04%**	**3.34**	**1.93**	**4.17%**	**4.24**

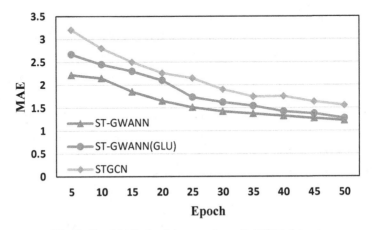

Fig. 6. Test MAE of training epochs on PeMSD4 dataset.

Effect of Graph Wavelet. Figure 6 shows the trend of Test MAE for the three methods over the training period in the PeMSD4 test set. With the increase of the training cycle, the prediction error of ST-GWANN decreases gradually, which makes it faster to stabilize, easier to converge, and achieve better prediction results than STGCN calculated by Fourier transform. Specifically, the localized convolution and spareness of the wavelet base greatly reduces the computational complexity of the graph neural network, which makes the computational process more efficient and ensures the accuracy of the model.

Effect of Attention Mechanism. To investigate the effects of attention mechanisms in ST-GWANN, we extracted a set of attention matrices from sensors. The left side of Fig. 7 shows the spatial location relationship of a road network with five sensors selected from PeMSD4. Figure 10 shows the attention matrix $R^{x \times y}$ between the sensors in the training set, and $R(x, y)$ represents the strength of correlation between the sensors x and y. For example, according to the fourth line, we can find that sensor 3 has a high correlation with sensor 4, which is reasonable because both sensor 3 and sensor 4 belong to the same road and are close to each other in real road network space, which indicates that it is effective to extract the correlation between hidden road network nodes by using the attention mechanism. Therefore, ST-GWANN accurately captures the spatial dependence of the road network, and the prediction performance in Table 1 and Table 2 is better than the model without an attention mechanism.

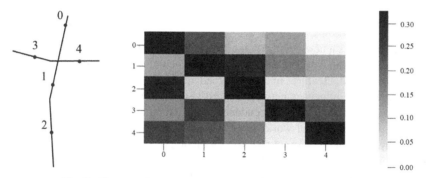

Fig. 7. The attention matrix obtained by graph attention module.

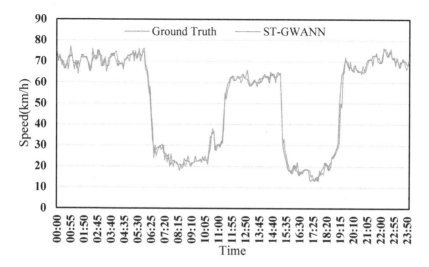

Fig. 8. Traffic prediction results of the dataset PeMSD4.

Visualization of Results. The visualization prediction results of ST-GWANN in a complete period on the PeMSD-4 dataset are shown in Fig. 8. It can be seen from the figure that the prediction results of ST-GWANN are very close to the ground truth. The results show that the spatial gated block can effectively use the spatial structure to predict traffic, and can capture the dynamic changes of traffic speed in the complex conditions of morning peak and evening peak. Compared with the ground truth, the predicted results fluctuate smoothly, which shows that GLU and the temporary transformer have achieved obvious effects on long-short term predictions. In summary, the proposed framework in this paper can make reliable predictions of complex traffic conditions.

6 Conclusion

This paper proposes a new deep learning framework called ST-GWANN for long-short term traffic prediction. We combine graph wavelet neural network and attention mechanism to extract spatial features in complex road networks. The attention mechanism can calculate different influences between nodes. Use GLU and the temporal transformer layer to obtain local and global dependence in the temporal dimension.

Experiments on two real-world datasets demonstrate the validity of the components proposed by the framework in actual traffic prediction tasks and obtain the best predictive results compared with the baseline model. In the future, we will optimize the network structure and incorporate external factors to improve the accuracy of traffic prediction.

Acknowledgment. This work is supported by National Key R&D Program of China (No.2017YFC0803300), the Beijing Natural Science Foundation (No.4192004), the National Natural Science of Foundation of China (No.61703013, 91646201, 62072016).

References

1. Jian, Y., Bingquan, F.: Synthesis of short-term traffic flow forecasting research progress. Urban Transpor. China **10**(6), 73–79 (2012)
2. Jia, Y., Wu, J., Xu, M.: Traffic flow prediction with rainfall impact using a deep learning method. J. Adv. Transpor. **2017**, 1–10 (2017)
3. Yu, B., Yin, H., Zhu, Z.: Spatio-temporal graph convolutional networks: a deep learning framework for traffic forecasting. In: International Joint Conference on Artificial Intelligence, pp. 3634–3640 (2017)
4. Wang, Z., Su, X., Ding, Z.: Long-term traffic prediction based on LSTM encoder-decoder architecture. IEEE Transac. Intellig. Transport. Syst. (2020)
5. Wu, Y., Tan, H.: Short-term traffic flow forecasting with spatial-temporal correlation in a hybrid deep learning framework. Comp. Sci. (2016)
6. Li, Y., Yu, R., Shahabi, C., Liu, Y.: Diffusion convolutional recurrent neural network: Data-driven Traffic Forecasting. (2017)
7. Xu, B., Shen, H., Cao, Q., Qiu, Y., Cheng, X.: Graph wavelet neural network. In: International conference for learning representations (ICLR 2019), pp.1–13. New Orleans, LA, USA (2019)
8. Vaswani, A., Shazeer, N., Parmar, N., Uszkoreit, J., Jones, L., Gomez, A.N., Polosukhin, I.: Attention is all you need. In: Advances in Neural Information Processing Systems (2017)
9. Smola, A.J., Schölkopf, B.: A tutorial on support vector regression. Stat. Comput. **14**(3), 199–222 (2004)
10. Kumar, S.V., Vanajakshi, L.: Short-term traffic flow prediction using seasonal ARIMA model with limited input data. Eur. Transp. Res. Rev. **7**(3), 1–9 (2015). https://doi.org/10.1007/s12 544-015-0170-8
11. Connor, J.T., Martin, R.D., Atlas, L.E.: Recurrent neural networks and robust time series prediction. IEEE Trans. Neural Netw. **5**(2), 240–254 (1994)
12. Hochreiter, S., Schmidhuber, J.: Long short-term memory. Neural Comput. **9**(8), 1735–1780 (1997)
13. Cho, K. et al.: Learning phrase representations using RNN encoder-decoder for statistical machine translation. Comp. Sci. (2014)
14. Parmar, N. et al.: Image transformer. In: International Conference on Machine Learning, pp. 4055–4064, PMLR (2018)

15. Zhang, J., Zheng, Y., Qi, D.: Deep spatio-temporal residual networks for citywide crowd flows prediction. In: Proceedings of the AAAI Conference on Artificial Intelligence, vol. 31, no. 1 (2017)
16. Bruna, J., Zaremba, W., Szlam, A., LeCun, Y.: Spectral networks and locally connected networks on graphs. In: Proceedings of International Conference on Learning Representations. (2014)
17. Kipf, T.N., Welling, M.: Semi-supervised classification with graph convolutional networks. In: International Conference on Learning Representations (ICLR) (2017)
18. Seo, Y., Defferrard, M., Vandergheynst, P., Bresson, X.: Structured sequence modeling with graph convolutional recurrent networks. In: International Conference on Neural Information Processing, pp. 362–373. Springer, Cham (2018)
19. Wu, Z., Pan, S., Long, G., Jiang, J., Zhang, C.: Graph wavenet for deep spatial-temporal graph modeling. In: International Joint Conference on Artificial Intelligence (2019)
20. Guo, S., Lin, Y., Feng, N., Song, C., Wan, H.: Attention based spatial-temporal graph convolutional networks for traffic flow forecasting. Proc. AAAI Conf. Artif. Intellig. **33**(01), 922–929 (2019)
21. Defferrard, M., Bresson, X., Vandergheynst, P.: Convolutional neural networks on graphs with fast localized spectral filtering (2016)
22. Wu, Z., Pan, S., Chen, F., Long, G., Zhang, C., Philip, S.Y.: A comprehensive survey on graph neural networks. In: IEEE Transactions on Neural Networks and Learning Systems (2020)
23. Veličković, P., Cucurull, G., Casanova, A., Romero, A., Lio, P., Bengio, Y.: Graph attention networks. In: International Conference on Learning Representations (2017)
24. Bahdanau, D., Cho, K., Bengio, Y.: Neural machine translation by jointly learning to align and translate. Comput. Sci. (2014)
25. Dauphin, Y.N., Fan, A., Auli, M., Grangier, D.: Language modeling with gated convolutional networks. In: International Conference on Machine Learning, pp. 933–941. PMLR (2017)
26. Amap Open Platform. https://lbs.amap.com/

Data Science

A Heterogeneity Urban Facilities Spatial Clustering Analysis Method Based on Data Field and Decision Graph

Lei Kang[1]([⊠]), Haiyan Liu[1], Xiaohui Chen[1], Weiying Cheng[2], Jing Li[1], Jia Li[1], and Chenye She[3]

[1] Information Engineering University, Zhengzhou 450001, China
[2] The 32137 Troop of PLA, Zhangjiakou 075001, China
[3] The 31675 Troop of PLA, Zhangjiakou 075001, China

1 Introduction

The clustering distribution characteristics of urban facilities is significant for capturing the current urban spatial structure. Correlation analysis of spatial clustering can deeply investigate the agglomeration mode of urban facilities. However, many existing clustering analyses commonly focus on the spatial features of entities abstracted as points without taking the attribute features into consideration, leading to the feature scale homogenization [1] and the inaccurate results inconsistent with some actual situations. More valuable spatial clustering analysis needs to take into account the differences of object attributes.

Data field model can describe the different roles of each data in the sample space for knowledge discovery, reflect the inherent attributes of data, and provide a link for the fusion of spatial information and attribute information [2]. In recent years, experts and scholars have made a lot of useful explorations on the combination of data field theory and spatial clustering analysis algorithm [3,4], proving the feasibility of applying data field theory to spatial clustering analysis.

In 2014, Alex Rodriguez [5] and others proposed decision graph to quickly identify cluster centers. In reference [7], a trajectory clustering method based on decision graph and data field (TCDGDF) is proposed by combining decision graph and data field. The data field is used to describe the interaction between trajectory points and determine the optimal parameters. Compared with the classical algorithm, the results show that this method can achieve trajectory clustering more effectively.

Based on the method of reference [7], this paper takes into account the attribute information of the sample objects, and uses the data field and decision graph to perform the spatial clustering analysis of urban facilities. The attribute characteristics of facilities are integrated into the spatial clustering analysis through the data field theory to show the interaction between heterogeneous urban facilities, and then the clustering center is selected and the clustering is completed by referring to the idea of decision graph.

G. Pan et al. (Eds.): SpatialDI 2021, LNCS 12753, pp. 103–109, 2021.
https://doi.org/10.1007/978-3-030-85462-1_8

2 Methods and Materials

2.1 Data Field

In physics, "field" is used to describe the state where interacting particles sometimes do not touch each other. In view of this, researchers [6] introduce the field model into the data space, and propose the data field to represent the interaction of data objects.

In the data field, the object radiates data energy to the sample space with its own center, and the field energy is isotropic and superimposed on each other. The potential of a point x refers to it receives the algebraic sum of all energy radiation. It can be defined as follows:

$$E(x) = \sum_{i=1}^{n} m_i \times K\left(\frac{\|x - x_i\|}{\sigma}\right) \tag{1}$$

In the function, m_i as the mass of data object i represents the feature weight of i. $\|x - x_i\|$ is the distance from x to the data object i. Impact factor σ controls the interaction scope among objects. $K(x)$ is the dimensional potential function, which determines the radiation mode of data in the sample space. Compared to other potential functions, Gaussian potential function showed as follows can better reveal the interaction among objects and is widely used.

$$E(x) = \sum_{i=1}^{n} m_i \times \exp\left(-\left(\frac{\|x - x_i\|}{\sigma}\right)^2\right) \tag{2}$$

2.2 Decision Graph

The main idea of decision graph is that the local density of cluster center is larger and more far away from other high density points. Through the decision graph, we can quickly find the data object with the highest density within the local scope, namely, the cluster center. Drawing decision graph involves two variables: the local density ρ_i and minimum distance δ_i. For data point i, the local density ρ_i is expressed as follows:

$$\rho_i = \sum_j \chi(d_{ij}, d_c) \tag{3}$$

In the function, $\chi(d_{ij}, d_c) = 1$ if $d_{ij} < d_c$ and $\chi(d_{ij}, d_c) = 0$ otherwise. d_c called the cutoff distance needs to be set.

Using cutoff function may result in the sudden change of ρ_i along with d_c. The description of local density is not accurate enough. In order to cope with the problem, replacing the truncation function with Gaussian kernel function enables the local density of sample data more stable dealing with the change of distance, and weaken the influence of parameter selection on the results. Gaussian kernel function is defined as follows:

$$\chi_{Gauss}(d_{ij}, d_c) = \exp\left(-\left(\frac{d_{ij}}{d_c}\right)^2\right) \tag{4}$$

The minimum distance δ_i based on the local density ρ_i represents the minimum distance between the data point i and other higher density points:

$$\delta_i = \min_{j:\rho_j > \rho_i} (d_{ij}) \tag{5}$$

As for the point with the highest density, δ_i represents the maximum distance from all other sample points: $\delta_i = max_j(d_{ij})$.

After calculating the local density ρ_i and minimum distance δ_i of the sample points, take ρ_i as the horizontal axis and δ_i as the vertical axis to draw the corresponding decision graph. As shown in Fig. 1(b), the points in the dotted line area have larger local density and minimum distance, and are selected as the clustering center, which is represented as the red marked points in Fig. 1(a).

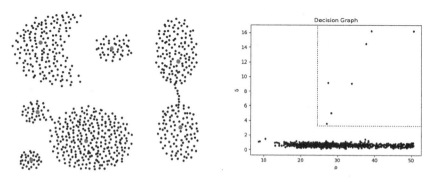

Fig. 1. Cluster center selection. (a) Sample data and cluster center, (b) decision graph

2.3 Potential Decision Graph

The potential decision graph is an improvement of the decision graph grounded on the data field theory. Based on the traditional decision graph, the horizontal axis of the potential decision graph replaces the local density ρ_i with the potential E_i. The equation is transformed from the Gaussian potential function Eq. (2):

$$E_i = \sum_{j=1}^{n} m_j \times \exp\left(-\left(\frac{d_{ij}}{d_c}\right)^2\right) \tag{6}$$

From the perspective of mathematics, without considering the quality, the Gaussian potential function is essentially a Gaussian kernel function Eq. (4), as shown in Fig. 2; in terms of the algorithm model, the meaning of the two is completely different. The potential deriving from Gaussian potential function is the superposition of objects' spatial features and attribute features, while the local density obtained by Gaussian kernel function only has the spatial meaning.

In many existing cases of data field potential combined with spatial clustering analysis [7], the data quality is usually not considered or set to 1. The field distribution of

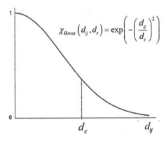 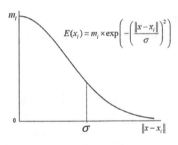

Fig. 2. Chart of function comparison. (a) Gaussian kernel function, (b) Gaussian potential function

a single data object is exactly the same, and the energy contribution to the surrounding data is only related to the distance. In essence, it is a clustering method using Gaussian kernel function to calculate the density, which just replaces the local density with the potential of data field in concept.

In this paper, data quality m_i represents the attribute characteristics of urban facilities and reflects the different effects of urban facilities. Compared with the traditional decision graph, due to the consideration of the quality differences, the potential decision graph can quickly discover the clustering centers with large local density and high quality so as to achieve the more realistic clustering results.

The next, we following the method of reference [5] classify the rest of objects into the nearest cluster of the higher potential objects until to the cluster centers.

3   Experiments and Analysis

Universities are important urban facilities that have received extensive attention. The comprehensive evaluation value of a number of indicators can reflect the overall situation and influence of universities. The research object of this paper is university facilities, and the university score is taken as the entity quality m_i to calculate the potential value of the object.

According to the traditional method, without referring to objects' attributes, the recognition results (see Fig. 3(a) and Table 1) show that there is no significant correlation between the cluster's ranking and the university's attributes, and the formation of cluster centers is only related to the geographical distribution. Compared with the results of considering the differences of University's attributes at the same time, the practical significance is not strong.

The main reason why Zhengzhou Normal University in Henan becomes a cluster center of the central region is that Henan ranks 3rd in terms of the quantity of universities among 31 provinces. But in fact, the average score of universities in Henan ranks 25th, which reveals the unusual slow development of higher education in Henan as a central province.

The results of this thesis are shown in Fig. 3(b) and Table 2. It can be found from the potential decision graph and cluster center information that the provinces and cities that the cluster center universities belong to have high-quality higher education resources, and the average score of the universities higher than other regions in the cluster is in positive correlation with their ranking. Except Shanghai Jiao Tong University, the cluster centers identified by potential energy decision graph are the universities with the highest score in the cluster. Shanghai Jiao Tong University ranking second in the cluster is the cluster center, which shows that the cluster center is not only determined by the its weight, but also interact with other objects in the local region, The results verify the effectiveness of spatial clustering analysis by integrating facility attribute features into data field model.

(a)   **(b)**

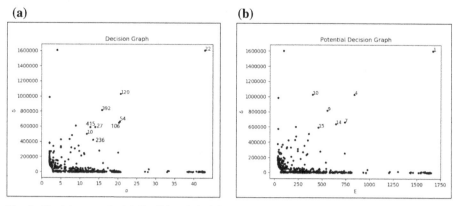

Fig. 3. Decision graph. (a) Traditional decision graph, (b) Potential decision graph. Numbers above the point represent the ranking of universities.

Table 1. Cluster centers of traditional method

University Name	Ranking in all universities	Ranking in clusters	Province	Ranking of Universities' quantity in province	Ranking of Universities' average score in province
Harbin Institute of Technology	10	1	Heilongjiang	16	8
Beijing Normal University	22	7	Beijing	1	1
University of Electronic Science and Technology	27	2	Sichuan	11	12

(*continued*)

Table 1. (*continued*)

University Name	Ranking in all universities	Ranking in clusters	Province	Ranking of Universities' quantity in province	Ranking of Universities' average score in province
Shenyang University	415	26	Liaoning	8	13
Zhongnan University of Economics and law	54	8	Hubei	5	4
Chang'an University	106	7	Shanxi	9	9
Shanghai Normal University	120	39	Shanghai	12	2
Guangdong University of Finance and Economics	392	17	Guangdong	6	6
Zhengzhou Normal University	236	14	Henan	3	25

Table 2. Cluster centers of this method

University	Ranking in all universities	Ranking in clusters	Province	Ranking of Universities' quantity in province	Ranking of Universities' average score in province
Tsinghua University	1	1	Beijing	1	1
Shanghai Jiao Tong University	4	2	Shanghai	12	2
Huazhong University of Science and Technology	7	1	Hubei	5	4

(*continued*)

Table 2. (*continued*)

University	Ranking in all universities	Ranking in clusters	Province	Ranking of Universities' quantity in province	Ranking of Universities' average score in province
Sun Yat-sen University	9	1	Guangdong	6	6
Harbin Institute of Technology	10	1	Heilongjiang	16	8
Xi'an Jiaotong University	14	1	Shanxi	9	9
Sichuan University	15	1	Sichuan	11	12

4 Conclusions

Given that the conventional spatial clustering analysis fails to describe sample objects' attribute characteristics, this paper proposes a new approach to reflect the heterogeneity of urban facilities. The main performances are as follows: the data field model maps the attribute value of urban facilities to the whole sample space so as to comprehensively describe the superposition effect of facilities' spatial characteristics and attribute characteristics; potential decision graph help to quickly get the maximum potential point in the local range and complete the clustering process. The experiment prove the practicality of the clustering results.

References

1. Wenhao, Y., Tinghua, A.: The visualization and analysis of POI features under network space supported by kernel density estimation. Acta Geodaet. Cartograph. Sin. **44**(01), 82–90 (2015)
2. Haijun, W., Deng, Y., Li, W., et al.: A C-means algorithm based on data field. Geomat. Inf. Sci. Wuhan Univ. **34**(05), 626–629 (2009)
3. Wang, S., Gan, W., Li, D., et al.: Data field for hierarchical clustering. Int. J. Data Warehouse. Min. **7**(4), 43–63 (2011)
4. Jing, Y., Jiawei, G., Jiye, L.: An improved DBSCAN clustering algorithm based on data field. J. Front. Comp. Sci. Technol. **6**(10), 903–911 (2012)
5. Rodriguez, A., Laio, A.: Clustering by fast search and find of density peaks. Science **344**(6191), 1492–1496 (2014)
6. Wenyan, G., Deyi, L., Jianmin, W.: An hierarchical clustering method based on data fields. Acta Electron. Sin. **2006**(02), 258–262 (2006)
7. Zhao, P., Qin, K., Ye, X., et al.: A trajectory clustering approach based on decision graph and data field for detecting hotspots. Int. J. Geogr. Inf. Syst. **31**(5–6), 1101–1127 (2016)

A Hybrid Algorithm for the Equal Districting Problem

Yunfeng Kong[(✉)] [iD]

Key Laboratory of Geospatial Technology for the Middle and Lower Yellow River Regions,
Ministry of Education, Henan University, Kaifeng 475000, China
yfkong@henu.edu.cn

Abstract. The equal districting problem (EDP) arises in applications such as
political redistricting, police patrol area delineation, sales territory design and
facility service area design. The important criteria for these problems are district equality, contiguity and compactness. A mixed integer linear programming
(MILP) model and a hybrid algorithm are proposed for the EDP. The hybrid
algorithm is designed by extending iterative local search (ILS) algorithm with
three schemes: population-based ILS, variable neighborhood descent (VND) local
search, and set partitioning. The performance of the algorithm was tested on five
areas. Experimentation showed that the instances could be solved effectively and
efficiently.

Keywords: Equal districting problem · Mixed integer linear programming ·
Hybrid algorithm

1 Introduction

Districting problems have been widely applied in geography, economics, environmental
science, politics, business, public service and many other areas. The equal districting
problem (EDP) arises in applications such as political redistricting, police patrol area
delineation, sales territory design and facility service area design. The basic criteria for
the problem are balance, compactness and contiguity [1, 2].

There are two general approaches to solve the EDP: exact methods and heuristics
[2, 3]. It is hard to solve the EDP by exact methods due to its computational complexity
[4]. Salazaraguilar et al. (2011) proposed a bi-objective programming model and solved
an instance with 150 units and six territories in four hours [5]. Duque et al. (2011) have
shown that only very small instances of the p-regions problem with 49 units could be
solved [6]. A new model proposed by Plane et al. (2019) was used to solve political
districting instances with 25–36 units in 6–53,632 s [7]. In practice, some districting
criteria were not fully considered in model formulations [3].

Early studies had also modelled the problem as a capacitated p-median facility location problem, and developed several location-based solution approaches [8–11]. It is not
necessary to determine district centers in solving the districting problem; however, districting measures such as contiguity and compactness rely on district centers [2]. Kong

© Springer Nature Switzerland AG 2021
G. Pan et al. (Eds.): SpatialDI 2021, LNCS 12753, pp. 110–120, 2021.
https://doi.org/10.1007/978-3-030-85462-1_9

et al. (2019) proposed a center-based model for solving the districting problem, in which the district centers were identified by a weighted K-medoids algorithm [12].

Three classes of heuristic methods were developed for districting problems: the local-search based heuristics such as greedy search, simulated annulling, tabu search, and old bachelor acceptance search [1, 13–17]; evolutionary methods such as evolutionary algorithm, genetic algorithm, and scatter search [18–21]; and swarm intelligence methods such as artificial bee colony [22]. However, the high-performance methods are always time consuming. For example, the PEAR algorithm [23] needs 3 h to solve an instance on a super computer with 131K processors.

This article aims to propose a high-performance hybrid algorithm for the EDP. The hybrid algorithm is designed by extending iterative local search (ILS) algorithm with three schemes: population-based ILS, variable neighborhood descent (VND) local search, and set partitioning. The performance of the algorithm was tested on five areas. Experimentation showed that the instances could be solved effectively and efficiently.

2 Mathematical Formulation

For a geographic area, let set $V = \{1, 2 \ldots n\}$ denote n basic areal units. Each unit i has an attribute value p_i. Let c_{ij} indicate whether units i and j share a border and N_i be a set of units that are adjacent to unit i ($N_i = \{j \mid c_{ij} = 1\}$). Let variables d_{ij} be the distance between units i and j. The EDP is to delineate a geographic area into K equal, compact and contiguous districts. A mixed integer linear programming (MILP) model is formulated as follows.

$$Min. \sum_{i \in V} \sum_{k \in V} p_i d_{ik} x_{ik} / \sum_{i \in V} p_i \tag{1}$$

$$S.T. \sum_{k \in V} x_{ik} = 1, \forall i \in V \tag{2}$$

$$(1 - \varepsilon)\overline{Q}y_k \leq \sum_{i \in V} p_i x_{ik} \leq (1 + \varepsilon)\overline{Q}y_k, \forall k \in V \tag{3}$$

$$\sum_{k \in V} y_k = K \tag{4}$$

$$f_{ijk} \leq (n - K)x_{ik}, \forall i \in V, j \in N_i, k \in V \tag{5}$$

$$f_{ijk} \leq (n - K)x_{jk}, \forall i \in V, j \in N_i, k \in V \tag{6}$$

$$\sum_{j \in N_i} f_{ijk} - \sum_{j \in N_i} f_{jik} \geq x_{ik}, \forall i \in V, k \in V, i \neq k \tag{7}$$

$$x_{ik} = \{0, 1\}, \forall i \in V, k \in V \tag{8}$$

$$y_k = \{0, 1\}, \forall k \in V \tag{9}$$

$$f_{ijk} \geq 0, \forall i \in V, j \in N_i, k \in V \tag{10}$$

The objective function (1) is to minimize the average travel distance from basic units to their district centers. Constraints (2) ensure that each unit i is assigned to only one district. Constraints (3) confirm that districts are almost equal with a predefined error $\varepsilon = 5\%$, e.g. ε. Note that $\overline{Q} = \sum_{i \in V} p_i / K$. Constraint (4) requests that the area is divided into K regions. The district contiguity is ensured by constraints (5), (6) and (7). It is a variant of the network-flow based model [6] to guarantee the district contiguity. Constraints (5) and (6) ensure that a flow may only be passed through neighboring units in the same district. If unit i does not serve as the center unit, constraints (7) state that one unit of flow must be created from this unit. Constraints (8), (9) and (10) define the decision variables.

3 Algorithm Design

A hybrid algorithm for the EDP was proposed based on iterative local search (ILS) algorithm. ILS is a simple, easy to implement, and quite effective metaheuristic for discrete optimization problems. It starts from an initial solution and iteratively improves the solution by local search and perturbation. Local search heuristic is used to intensively explore the solution space. However, the iterative search may easily get trapped in local optima that are far away from the global optimum. ILS escapes from local optima by applying perturbations to the current local minimum.

In this article, the standard ILS algorithm was enhanced by three schemes: population-based search, variable neighborhood descent (VND) search, and set partitioning. Given parameters such as population size (*psize*), perturbation strength (*strength*), number of consecutive loops that the best solution is not updated (*mloops*), time limit for set partitioning (*tlimit*), the hybrid algorithm is outlined as follows:

```
1.  P = GenerateInitialSolutions(psize);
2.  pool=null;
3.  s_best=Best(P);
4.  notImpr=0;
5.  While notImpr < mloops:
6.      Select a solution s from P randomly;
7.      s'=Perturbation(s,strength);
8.      s"=VNDsearch(s');
9.      s*=updateCenetrs(s")
10.     If f(s*)<f(s_best):  s_best=s*,notImpr=0;
11.         else: notImpr+=1;
12.     pool=UpdateAreaPool(pool,s*);
13.     P=UpdatePopulation(P,s*);
14. s=SetPartitioning(pool,t);
15. Output s.
```

The hybrid algorithm maintains a number of candidate solutions. In step (1), initial solutions for the EDP are constructed by a weighted K-medoids algorithm [12]. In step (6), a solution is randomly selected from the population as the incumbent solution. After solution perturbation and local search, in step (13), the population is updated according to the solution objectives and the similarities among the solutions. The elite and diverse solutions are selected by the algorithm. As a result, the population-based extension of ILS may enhance the diversification of local search.

The design of local search operators in optimization algorithm is critical for repeatedly improving the incumbent solution. The one-unit shift is a widely used operator that moves a boundary unit to its neighboring district while maintaining the connectivity of its original district. Butsch et al. introduced three operators for districting problems: shift, double shift, and swap [24]. In the proposed algorithm, two local search operators are used to improve the solutions: one-unit shift and two-unit shift. Instead of simple local search in ILS, VND search is used in step (8). VND is repeatedly exploring different neighborhood structures by the change of neighborhoods until the solution cannot be improvement. Using VND search, ILS algorithm can easily find solutions in local optimum.

One-unit shift and two-unit shift operators are implemented as follow. For each boundary unit i, the algorithm tries to move the boundary unit from its original area to one of its neighboring areas. The move is accepted in cases that the original area is connective, and the new solution is better than the incumbent solution. The two-unit shift moves boundary unit i to one of its neighboring areas k, and at the same time moves boundary unit j in area k to one of its neighboring areas. In other words, one unit is moved into area k, and another unit in area k is moved out. Differing from the one-unit shift that involves two areas, a two-unit shift usually involves three areas. Swapping two units between two adjacent areas is a special case that involves two areas. The two local search operators have different search space and computational complexity: $O(Kn)$ and $O(Kn^2)$. However, the number of possible moves is generally much fewer, because only the boundary units can be moved to their neighboring areas, and there are only a few neighboring areas available for a boundary unit.

Perturbation is one of the key components of ILS algorithm. Using local search operators, the solution may easily reach to local optima. A ruin and recreate procedure is useful to perturb the EDP solution from local optima. There are multiple methods to ruin the solution such as deleting some boundary units, deleting all the units in some districts, and deleting some units in a connective region. The ruined solution must be repaired to be a feasible solution.

In step (9), the district centers are updated by selecting the best unit in each district. In any district of the current solution, if the center can be updated, the objective will be decreased.

In step (14), the best solution might be improved by a set partitioning procedure. In each ILS loop, all the districts in solution $s*$ are recorded in the list *pool*. Each district record includes the units in the district and its objective value. At the end of ILS, thousands or more districts will be recorded in the pool. A set partitioning problem (SPP) model could be used to select a better solution from the candidate service areas. Let Ω be the set of districts identified in step (9), U_i and O_i be the set of basic units in each district i and its objective, respectively. A SPP model is used to select K districts from Ω. Let Ω_j be a subset of Ω, $\Omega_j = \{i|i \in \Omega, j \in U_i\}$, the SPP model is formulated as follows:

$$Min. \sum_{i \in \Omega} O_i x_i \tag{11}$$

$$S.T. \sum_{i \in \Omega_j} x_i = 1, \ \forall j \in V \tag{12}$$

$$\sum_{i \in \Omega} x_i = K \tag{13}$$

$$x_i = \{0, 1\}, \forall i \in \Omega \tag{14}$$

4   Experimentation

Five geographic areas (Fig. 1) are selected to test the EDP. The attribute value p_i for each spatial unit are shown as grey circles in the maps. The main features of the areas are summarized in Table 1. Areas HD, ZY, GY, GY2 and HD are different in terms of the area, the geographic background and the number of spatial units.

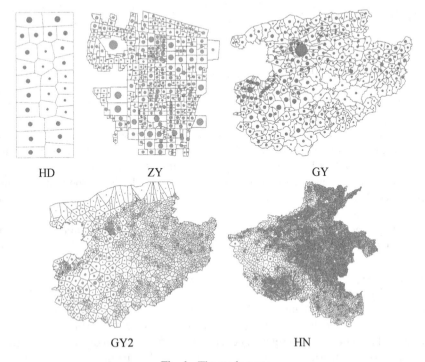

HD   ZY   GY

GY2   HN

Fig. 1. The study areas

Table 1. Main features of the study areas

Area	HD	GY	ZY	GY2	HN
Geography	Residence community	County	Urban	County	Province
No. of basic units	28	297	324	1,276	2,145
Sum of attribute value p_i	2,420	36,824	3,873	819,812	96,918,036

Table 2. Districting results from the areas HD, GY and ZY

Area	K	ε	CPLEX				Hybrid		
			LB	UB	Gap	Time/s	Gap	Dev	Time/s
HD	2	5%	0.099	0.099	Opt.	0.99	0.00%	0.00%	0.32
HD	3	5%	0.084	0.084	Opt.	1.20	0.00%	0.00%	0.44
HD	4	5%	0.071	0.071	Opt.	1.14	0.00%	0.00%	0.52
HD	5	5%	0.060	0.060	Opt.	1.31	0.00%	0.00%	0.62
GY	2	5%	8.149	8.149	Opt.	6,098.57	0.00%	0.00%	6.20
GY	3	5%	6.085	6.085	Opt.	3,990.47	0.04%	0.06%	6.18
GY	4	5%	5.135	5.135	Opt.	11,145.66	1.19%	0.03%	7.51
GY	5	5%	4.616	4.616	Opt.	7,588.53	0.16%	0.07%	9.92
GY	6	5%	4.185	4.185	Opt.	11,303.26	0.24%	0.37%	6.67
GY	7	5%	3.751	3.751	Opt.	3,576.36	0.00%	0.00%	6.17
GY	8	5%	3.529	3.529	Opt.	5,390.92	0.22%	0.09%	6.92
GY	10	5%	3.120	3.120	Opt.	4,823.14	0.00%	0.01%	7.18
ZY	2	5%	0.979	0.979	Opt.	3,480.72	0.00%	0.00%	5.24
ZY	3	5%	0.831	0.831	Opt.	3,796.54	0.00%	0.00%	5.75
ZY	4	5%	0.717	0.717	Opt.	4,292.85	0.01%	0.04%	5.84
ZY	5	5%	0.624	0.624	Opt.	4,758.05	0.05%	0.03%	5.80
ZY	6	5%	0.562	0.562	Opt.	5,811.00	0.08%	0.02%	6.04
ZY	7	5%	0.513	0.513	Opt.	5,927.70	0.01%	0.01%	6.00
ZY	8	5%	0.477	0.477	Opt.	4,544.64	0.02%	0.04%	7.32
ZY	10	5%	0.425	0.425	Opt.	4,973.77	0.02%	0.01%	7.03
ZY	15	5%	0.351	0.351	Opt.	4,973.77	0.17%	0.10%	9.84

The proposed algorithm was implemented by using the Python programming language, and was executed in PyPy 7.0, a fast and compliant implementation of the Python language (www.pypy.org). The variables p_i, c_{ij} and d_{ij} for each area were prepared in ArcGIS 10 (www.esri.com). All the software ran on a desktop computer with Intel Core I7–6700 CPU 3.40-GHz, 8 GB RAM and the Windows 10 operating system.

The problem instances were solved by both exact and heuristic methods, respectively. First, each instance model was solve by CPLEX Optimizer 12.6. A solution with lower bound (LB), upper bound (UB), MIPGap between LB and UB (Gap) and computation time (Time) could be obtained from CPLEX optimizer. Note that only the instances on small and medium areas HD, GY and ZY can be solved by CPLEX. Second, each instance was repeatedly solved by the hybrid algorithm for ten times. The parameters were set as follows: $\varepsilon = 5\%$, $psize = 10$, $strength = 5\%$, $mloops = 100$, and $tlimit = 100$ s. The heuristic solutions are evaluated in terms of the average gap to lower bound (Gap), the standard deviation among 10 solutions (Dev) and the average computation time (Time).

Districting results from areas HD, GY and ZY are summarized in Table 2. The solution results show that all the instances can be optimally solved by CPLEX optimizer; and the solutions from the hybrid algorithm approximate to the optimal solutions or the lower bounds with small gaps ranging from 0.00% to 1.19%. Part of solutions are illustrated in Fig. 2, 3 and 4.

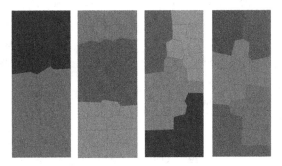

Fig. 2. Districting maps of the area HD

Fig. 3. Districting maps of the area GY

Districting results from areas GY2 and HN are summarized in Table 3. Due to large numbers of basic units in the two areas, the CPLEX failed to solve the instance models with memory overflow errors. Among 10 solutions for each instance, the best, average and worst objective are shown in columns Min, Avg and Max, respectively. The standard

Fig. 4. Districting maps of the area ZY

deviation of objectives and the average computing time are illustrated in columns Dev and Time, respectively. The results show that the large problems could be efficiently solved in 2–9 min with very small deviations (<0.18%). Part of districting maps are illustrated in Fig. 5 and 6.

Table 3. Districting results from areas GY2 and HN

Area	K	ε	Min	Avg	Max	Dev	Time/s
GY2	5	5%	4.439	4.442	4.443	0.03%	115.78
GY2	10	5%	3.108	3.109	3.110	0.03%	123.96
GY2	15	5%	2.544	2.548	2.551	0.08%	268.82
GY2	20	5%	2.174	2.177	2.181	0.11%	165.05
HN	5	5%	69.762	69.766	69.769	0.00%	302.46
HN	10	5%	46.548	46.560	46.586	0.02%	330.24
HN	15	5%	37.562	37.571	37.581	0.02%	546.56
HN	20	5%	31.781	31.795	31.810	0.03%	404.85
HN	25	5%	28.291	28.307	28.327	0.04%	539.99
HN	30	5%	25.854	25.864	25.879	0.03%	340.52
HN	35	5%	23.764	23.798	23.832	0.09%	323.62
HN	40	5%	22.244	22.296	22.333	0.13%	498.67
HN	45	5%	20.760	20.802	20.860	0.18%	388.39
HN	50	5%	19.541	19.590	19.612	0.12%	410.77

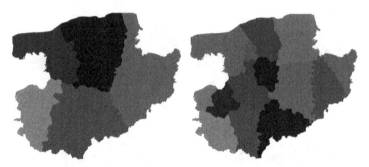

Fig. 5. Districting maps of area GY2

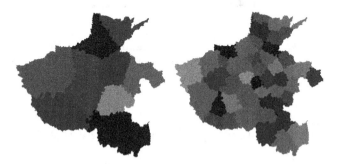

Fig. 6. Districting maps of area HN

5   Conclusion

A mathematical model and a hybrid algorithm were proposed for the EDP in this paper. The MILP model is different from existing models. First, it is a center-based model, and thus it is easy to formulate an objective function on district compactness. Second, the constraints on district contiguity is simplified by the center units of districts. Third, the EDP is defined as an equally capacitated p-median problem with constraints of district contiguity. As a result of the center-based formulation, the proposed model could optimally solve EDP instances with 324 basic units, while the districting models in [6] and [7] could only solve small instances with 16–49 or 25–36 units, respectively.

The proposed hybrid algorithm is different from existing algorithms in some aspects. The perturbation in ILS algorithm plays an important role in escaping from local optima and an appropriate perturbation strength could guide the ILS search approaching the global optimum. The algorithm was designed based on the standard ILS algorithm, and further enhanced by three schemes: VND local search, population-based ILS, and set partitioning. The experimentation showed that the EDP instances could be solved effectively and efficiently. The solutions found by the hybrid algorithm approximate optimal solutions or the lower bounds with very small gaps, and are also robust with deviations less than 0.20%.

The EDP has potential applications for emergency management in the COVID-19 pandemic. In case of the lockdown in some cities in China over COVID-19 outbreak, the residents were banned from leaving their residence communities. It is necessary to provide services such as the medicine distribution, the supply of daily necessities and emergent large-scale nucleic acid detection of COVID-19 for all residents. The districts and their centers of the EDP provided by the proposed method are very useful in such emergency service planning. It is easy to create equal districts in a region according to its population distribution. The residents in each district could be served by the facility located at the district center. The number of districts could be estimated by the total service demand, the service capacity per facility, and human resources. There are three advantages to provide such emergent service plans. First, a service plan could be prepared quickly in a short time, since these is no need to prepare the candidate locations for setting up facilities. Preparing the candidate locations is usually time-consuming for planning of schools, healthcare centers and many other public services. Second, the spatial accessibility of services is ensured by the EDP criterion of district compactness. Third, the service duties could be equally assigned to each staff or staff group. The author believes that some lockdown emergency response plans could be prepared efficiently and effectively based on the EDP solutions.

Foundation Support: The National Natural Science Foundation of China, No. 41871307.

References

1. Ricca, F., Simeone, B.: Local search algorithms for political districting. Eur. J. Oper. Res. **189**(3), 1409–1426 (2008)
2. Kalcsics, J.: Districting problems. In: Laporte, G., Nickel, S., da Gama, F.S. (eds.) Location Science, pp. 595–622. Springer, Cham (2015). https://doi.org/10.1007/978-3-319-13111-5_23
3. Ricca, F., Scozzari, A., Simeone, B., et al.: Political districting: from classical models to recent approaches. Q. J. Oper., Res. **9**(3), 223–254 (2011)
4. Keane, M.C.: The size of the region-building problem. Environ. Plan. A **7**(5), 575–577 (1975)
5. Salazaraguilar, M.A., Riosmercado, R.Z., Gonzalezverlarde, J.L.: A bi-objective programming model for designing compact and balanced territories in commercial districting. Transp. Res. Part C: Emerg. Technol. **19**(5), 885–895 (2011)
6. Duque, J.C., Church, R.L., Middleton, R.S., et al.: The p-Regions problem. Geogr. Anal. **43**(1), 104–126 (2011)
7. Plane, D.A., Tong, D., Lei, T.: Inter-person separation: a new objective standard for evaluating the spatial fairness of political redistricting plans. Geogr. Anal. **51**, 251–279 (2019)
8. Hess, S.W., Weaver, J.B., Siegfeldt, H.J., et al.: Nonpartisan political redistricting by computer. Oper. Res. **13**(6), 998–1006 (1965)
9. Hojati, M.: Optimal political districting. Comput. Oper. Res. **23**(12), 1147–1161 (1996)
10. George, J.A., Lamar, B.W., Wallace, C.A.: Political district determination using large-scale network optimization. Socio-Econ. Plann. Sci. **31**(1), 11–28 (1997)
11. Nemoto, T., Hotta, K.: Modelling and solution of the problem of optimal electoral districting. Commun. Oper. Res. Jpn **48**(4), 300–306 (2003). (in Japanese)
12. Kong, Y., Zhu, Y., Wang, Y.: A center-based modeling approach to solve the districting problem. Int. J. Geogr. Inf. Sci. **33**(2), 368–384 (2019)
13. Garfinkel, R., Nemhauser, G.L.: Optimal political districting by implicit enumeration techniques. Manage. Sci. **16**(8), 495–508 (1970)

14. Nygreen, B.: European assembly constituencies for Wales - comparing of methods for solving a political districting problem. Math. Program. **42**(1), 159–169 (1988)
15. Mehrotra, A., Johnson, E.L., Nemhauser, G.L.: An optimization based heuristic for political districting. Manage. Sci. **44**(8), 1100–1114 (1998)
16. Damico, S.J., Wang, S., Batta, R., et al.: A simulated annealing approach to police district design. Comput. Oper. Res. **29**(6), 667–684 (2002)
17. Bozkaya, B., Erkut, E., Laporte, G., et al.: A tabu search heuristic and adaptive memory procedure for political districting. Eur. J. Oper. Res. **144**(1), 12–26 (2003)
18. Wei, B.C., Chai, W.Y.: A multiobjective hybrid metaheuristic approach for GIS-based spatial zoning model. J. Math. Model. Algorithms **3**(3), 245–261 (2004)
19. Bacao, F., Lobo, V., Painho, M., et al.: Applying genetic algorithms to zone design. Soft. Comput. **9**(5), 341–348 (2005)
20. Chou, C., Chu, Y., Li, S., et al.: Evolutionary strategy for political districting problem using genetic algorithm. In: International Conference on Conceptual Structures, (2007)
21. Salazar-Aguilar, M.A., Rios-Mercado, R.Z., Gonzalez-Velarde, J.L., et al.: Multiobjective scatter search for a commercial territory design problem. Ann. Oper. Res. **199**(1), 343–360 (2012)
22. Rincón-García, E.-A., et al.: ABC, a viable algorithm for the political districting problem. In: Gil-Aluja, J., Terceño-Gómez, A., Ferrer-Comalat, J.C., Merigó-Lindahl, J.M., Linares-Mustarós, S. (eds.) Scientific Methods for the Treatment of Uncertainty in Social Sciences. AISC, vol. 377, pp. 269–278. Springer, Cham (2015). https://doi.org/10.1007/978-3-319-19704-3_22
23. Liu, Y., Choc, W.K., Wang, S.: PEAR: a massively parallel evolutionary computation approach for political redistricting optimization and analysis. Swarm Evol. Comput. **30**, 78–92 (2016)
24. Butsch, A., Kalcsics, J., Laporte, G.: Districting for arc routing. INFORMS J. Comput. **26**(4), 809–824 (2014)

An Integrated Correction Algorithm for Multi-Node Data from the Hydro-Meteorological Monitoring System in the South China Sea

Wuyang Chen[1,2], Junmin Li[1,2,3(✉)] (iD), Junliang Liu[1], Bo Li[1,2,4], Ping Shi[1,3,4], Huanlin Xing[1], and Xiaomin Long[1]

[1] State Key Laboratory of Tropical Oceanography, South China Sea Institute of Oceanology, Chinese Academy of Sciences, Guangzhou, China
jli@scsio.ac.cn
[2] Key Laboratory of Science and Technology on Operational Oceanography, Chinese Academy of Sciences, Guangzhou, China
[3] Southern Marine Science and Engineering Guangdong Laboratory (Guangzhou), Guangzhou, China
[4] Guangdong Key Laboratory of Ocean Remote Sensing, Guangzhou, China

Abstract. The data obtained by ocean monitoring systems have the characteristics of multi-source, heterogeneous, large quantity, and high timeliness, so it is necessary to formulate a correction algorithm to integrate into the software system of receiving server. An algorithm for multi-node and real-time data correction is developed for the hydro-meteorological (current, wave, tide level, wind speed, etc.) monitoring system including more than twenty nodes deployed in the South China Sea. The algorithm integrates different types of data with different protocols obtained by multi-node and processes the data at the code level. In the initial character string correction subroutine, the key characters are searched by regular expression, and the corresponding character is added to modify the starting string. In the data array value correction subroutine, the parameters are extracted and arranged, and the suspicious data is replaced, revised, or marked according to the parameter formats and value ranges. The results show that the algorithm has good universality to correct most of the errors in the data of various observation instruments. It greatly improves the amount of available data and the receiving rate for different types of monitoring nodes.

Keywords: Monitoring system · Data correction · The South China Sea · Marine hydrology · Marine meteorology

1 Introduction

Accurate and efficient marine hydrological (e.g., current, wave, tide level, etc.) and meteorological (e.g., wind speed, temperature, humidity, pressure, etc.) data are of great importance in the fields of disaster prevention, trade, shipping, energy exploitation,

© Springer Nature Switzerland AG 2021
G. Pan et al. (Eds.): SpatialDI 2021, LNCS 12753, pp. 121–126, 2021.
https://doi.org/10.1007/978-3-030-85462-1_10

and engineering construction in the South China Sea (SCS). By deploying nodes at several key stations, a hydro-meteorological monitoring system can obtain multi-source, heterogeneous, and large amounts of data that transfer to the laboratory in real-time [1–3]. Because the marine environment is complex and sometimes terrible, the problems of irregular errors such as random garbled code and byte missing would arise in the data transmission process. To ensure the availability and timeliness of the data, it is necessary for the received data to synchronously enter preprocessing steps [3]. However, most related software mainly focuses on data cleaning, quality control, or the terminal application such as data visualization, and user interface [4, 5]. Although recent studies have tried to retrieve the monitoring data lost during transmission, by using, for example, intelligent optimization estimation algorithm [6]; there are still few studies that focus on data repairing at the coding level to restore the disordered or missing character to effective data. In this paper, a synchronous correction algorithm at the coding level was developed for multi-node and real-time data from the hydro-meteorological monitoring system of more than twenty nodes in the SCS.

2 Hydro-meteorological Monitoring System

The hydro-meteorological monitoring systems were planned to deploy at eight typical stations, with a total of twenty-five monitoring nodes in the SCS. The locations of the station are shown in Fig. 1. From 2018 to 2020, thirteen monitoring nodes have been deployed and run at six stations. The monitoring nodes include two types of platforms: mooring buoy and bottom-mounted wave and current profiler. The appearances of the platforms are shown in Fig. 1. The mooring buoy integrates a portable weather station, a wave sensor, and a current profiler. The bottom-mounted profiler is connected to the shore-based platform through cables to send back data [5]. The system monitors wave parameters (such as significant height, mean 1/3 height, mean 1/10 height, maximum height, mean height, mean period, peak period, mean zero-crossing period, mean 1/3 period, mean 1/10 period, maximum period, peak direction, directional spread, and mean direction), wave frequency and directional spectra, current velocity profiles, sea surface parameters (such as water level, current velocity, current direction, temperature, and sound velocity), and meteorology parameters (wind speed, wind direction, air temperature, air pressure, relative humidity, and dew point temperature). The data volume is about 16 kb/hour for the mooring buoy and 8 kb/hour for the bottom-mounted profiler. The data have been used in resource development and wave forecasting [7, 8], and provide information for multi-user in sea area management, engineering design and maintenance, and sustainable development of the environment.

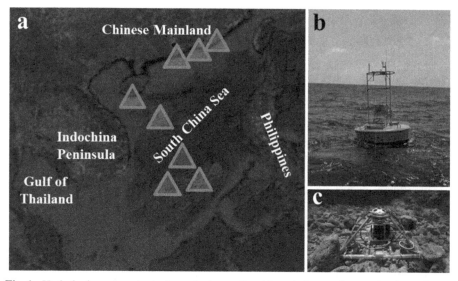

Fig. 1. Hydrologic-meteorological monitoring station (a), and photos of a mooring buoy (b) and bottom-mounted wave and current profiler (c).

3   Correction Algorithm of Monitoring Data

Due to the characteristics of multi-source, heterogeneous, large amount, and high time-liness of the data, it is necessary to formulate a synchronous correction algorithm integrated into the software system of the receiving server. By analyzing the data, the errors in the data can be divided into eight types: starting keyword missing, ending keyword missing, abnormal ending keyword, the abnormal character in rows, the repeated character in rows, mixing of different elements due to missing codes, mixing of different elements and mixed with abnormal characters, and error codes in little-endian. Corresponding correction methods for the transmission data code are formulated and then integrated into the calculation process of the system (Fig. 2). The multi-node data of different types and protocols are firstly integrated into the process (step 2). The subsequent operation mainly includes two subroutines: the initial string correction (step 4) and the array value correction (step 7). The former mainly searches for key characters through regular expressions and adds corresponding starting strings to modify the starting strings; in the latter, the effective fragments are marked to make the codes aligned to the correct parameter column, according to the value range and arrangement law. The algorithm has good universality and can complete synchronous correction processing for the data sent back by various monitoring platforms and equipment.

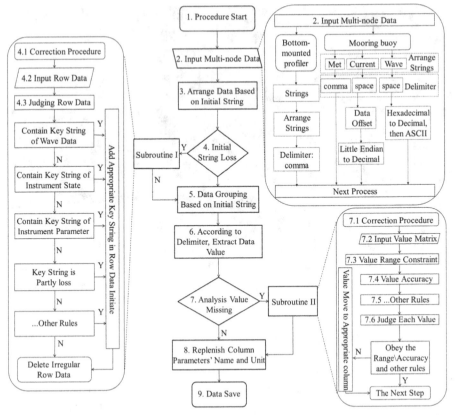

Fig. 2. Flow chart of the integrated correction algorithm for multi-node data from the hydro-meteorological monitoring system.

4   Correction Results

Based on the mooring buoy data of 33 nodes × month and the bottom-mounted profiler data of 104 nodes×month from 2018 to 2020. The effective data amount and reception rate of the two types of monitoring nodes before and after the implementation of the correction algorithm are compared. As can be seen from Fig. 3, after the correction, the effective data of wave, current, and meteorological parameters measured by the mooring buoy are increased by 11 kb/month, 53 kb/month, and 47 kb/month respectively, and the data reception rate is increased by 8–16%; The effective data volume of wave and current parameters measured by the bottom-mounted profiler is increased by 6 kb/month and 27 kb/month respectively, and the data receiving rate is increased by 4–6%. For the key parameters, such as significant wave height, current velocity, and wind speed measured by the buoy, the data volume increased by 1.9 group/day, 3.5 group/day, and 4.7 group/hour respectively after the correction; the significant wave height, current velocity, and water level measured by the bottom-mounted profiler increased by 5.9 group/day, 1.8 group/day, and 3.3 group/day respectively. These results show that the proposed algorithm can effectively and synchronously improve the effective ratio for the multi-node monitoring data.

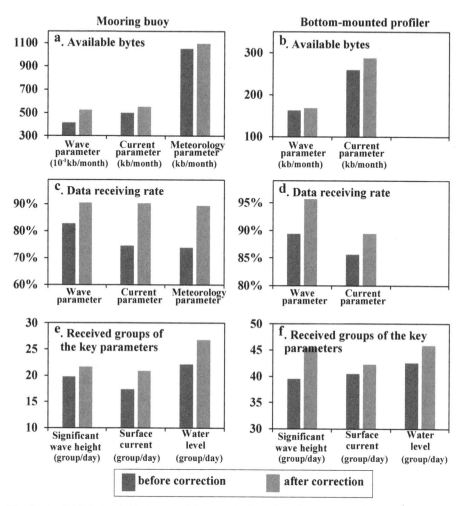

Fig. 3. Available bytes (a,b), data receiving rate (c,d), and received groups of the key parameters (e,f) before and after correction for mooring buoys (a,c,d) and bottom-mounted wave and current profilers (b,d,f).

5   Conclusion

Combined with the data structure of the hydro-meteorological real-time monitoring system deployed in the SCS, an integrated correction algorithm for multi-node monitoring data is constructed based on the transmission coding level. From 2018 to 2020, thirteen monitoring nodes have been deployed and run successively. With the help of the algorithm, the code errors that arise in the transmission process are quickly recovered, which significantly improves the amount of effective data received. The observation data has provided a lot of basic information support for ocean engineering and numerical simulation verification. In the future, considering the continuous development of diversified ocean observation platforms, it is necessary to develop a new generation of data correction algorithm combined with machine learning technology.

Acknowledgements. This study is supported by the Key Special Project for Introduced Talents Team of Southern Marine Science and Engineering Guangdong Laboratory (Guangzhou) (GML2019ZD0303), the CAS Key Technology Talent Program (173059000000160006), the National Key Research and Development Program of China (2017YFC1405404), the Strategic Priority Research Program of the Chinese Academy of Sciences (XDA13030304), and the Guangzhou Basic and Applied Basic Research Project (202102020464). The analysis of the algorithm in this study is supported by the High Performance Computing Division in the South China Sea Institute of Oceanology.

References

1. Li, J., et al.: Development of international marine observation system and construction of deep-sea station in China. J. Trop. Oceanogr. **31**(2), 123–133 (2012) (in Chinese)
2. Long, X., Wang, S., Cai, S., Chen, J.: SZS3–1 type wave-tide gauge. J. Trop. Oceanogr. **24**(3), 81–85 (2005). (in Chinese)
3. Xu, G., Shi, Y., Sun, X., Shen, W.: Internet of things in marine environment monitoring: a review. Sensors. **19**(7), 1711 (2019)
4. Ionel, R., Pitulice, L., Vasiu, G., Mischie, S., Spiridon, O.B.: Implementation of a GPRS based remote water quality analysis instrumentation. Measurement **65**, 81–93 (2015)
5. Li, B., Chen, W., Liu, J., Li, J., Wang, S., Xing, H.: Construction and application of nearshore hydrodynamic monitoring system for uninhabited islands. J. Coast. Res. **99**(SI), 131–136 (2020)
6. Chen, X., Jin, Z., Lin, G., Yang, X.: Using node clustering and genetic programming to estimate missing data for marine environmental monitoring. J. Coast. Res. **73**(SI), 711–719 (2015)
7. Li, B., et al.: Application of artificial neural network to numerical wave simulation in the coastal region of island. J. Xiamen Univ. (Nat. Sci.) **59**(3), 420–427 (2020). (in Chinese)
8. Li, B., et al.: Wave energy resources in nearshore area of Dongluo Island, Sanya. IOP Conf. Ser.: Earth. Environ. Sci. **467**, 012079 (2020)

Comparison and Analysis of Hexagonal Discrete Global Grid Coding

Long Zhao[1,2], GuoQing Li[1,2], XiaoChuang Yao[3(✉)], QianQian Cao[1,2], and Yue Ma[1,2]

[1] Aerospace Information Research Institute, Chinese Academy of Sciences,
Beijing 100094, China
[2] School of Electronic, Electrical and Communication Engineering, University of Chinese
Academy of Sciences, Beijing 100049, China
[3] College of Land Science and Technology, China Agricultural University,
Beijing 100083, China
yxc@cau.edu.cn

Abstract. The Discrete Global Grid System (DGGS) have characteristics of discrete, hierarchical and global continuity, which not only meets the requirements of data discretization and parallel processing in the era of big data, but also get rid of the shackles of traditional map projection. It also have potential capacity to handle global multi-resolution massive spatial data. Hexagonal Discrete Global Grid has good geometric properties and has become a hot spot in the research and application of DGGS. However, how to establish an efficient coding operation scheme on the sphere is a challenge of the current research and application. According to the structural characteristics of the hexagonal division, this paper constructs the calculations of hierarchical coding, integer coordinate coding and filling curve coding, compares their advantages and disadvantages. The test results show that integer coordinate coding has the highest efficiency in addition operation and adjacent cell search, but it is described on two dimensions, which is not conducive to grid identification and storage In conclusion, operation efficiency of filling curve coding is higher to hierarchical coding, and it is easy to represent and store, which is more suitable for DGGS coding and coding operation.

Keywords: Discrete Global Grid System · Hexagon · Code operation

1 Introduction

A Discrete Global Grid System (DGGS) is a spatial reference system that uses a hierarchical tessellation of cells to partition and address the globe [1]. The coding operation is the core of the DGGS, which supports the rapid indexing of the spatial data of the entire system and the efficient calculation of application analysis [2]. According to the coding principle, the encoding scheme is roughly divided into three types: hierarchical coding, integer coordinate coding and filling curve coding [3]. Among them, the hierarchical coding operation includes QTM [4] and PYXIS [5, 6], HQBS [7], HLQT [8, 9], etc. Integer coordinate coding uses two-dimensional and three-dimensional integer coding to describe the center of gravity of the unit and establish an index [5, 10, 11]. The filling

© Springer Nature Switzerland AG 2021
G. Pan et al. (Eds.): SpatialDI 2021, LNCS 12753, pp. 127–133, 2021.
https://doi.org/10.1007/978-3-030-85462-1_11

curve coding uses Morton code to code the four-hole diamond grid on the icosahedron and the regular octahedron [12, 13].

Different coding methods have their own advantages and disadvantages in terms of computational efficiency and coding attributes. Therefore, comparative studies are needed to provide supports for the further research and application of DGGS. Compared with triangles and quadrilaterals, hexagons have more desirable geometrical and spatial characteristics. Based on the Aperture 4 Hexagonal Discrete Global Grid System, this paper constructs three kinds of codes: hierarchical coding, integer coordinate coding, and filling curve coding, establishes code addition operation and adjacent cell search algorithms, and compares the efficiency of the three coding operations through experiments.

2   Icosahedral Hexagonal DGGS

The hexagonal DGGS usually make use of platonic solids as initial polyhedrons. Among them, the icosahedron has smallest distortion because of the largest number of triangular. The discretization on the icosahedron is mainly to expand the icosahedron into twenty triangular and establish an initial division structure on each triangular. Finally, as shown in Fig. 1, 20 face hexagons and 12 vertice hexagons are constructed based on the initial discretization. So it is necessary to establish the discretization structure and encoding scheme on the hexagonal plane, which could be expanded to the whole icosahedron and sphere.

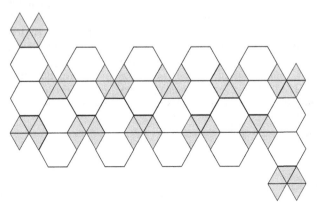

Fig. 1. The expansion of icosahedron

3   Coding Scheme and Operation of Hexagonal Discrete Global Grid System

Nearly all existing hexagonal DGGSs are built from planar coding scheme to polyhedron, and then projected on the sphere. This paper constructs these three kinds of coding scheme and operation on the planar aperture 4 hexagonal DGGS.

3.1 Hierarchical Coding

The hierarchical coding uses the hierarchical structure generated by the recursive division of hexagons on the plane to identify the hexagonal grids of each level. As shown in Fig. 2, this paper constructs a hexagonal "quadtree" coding structure on the plane [9]. For the seven hexagons in the first layer, the coding of scheme is identified as {0, 1, 2, 3, 4, 5, 6}. From the second layer, four sub-hexagons are divided from initial grid and coded as {0, 1, 2, 3}.

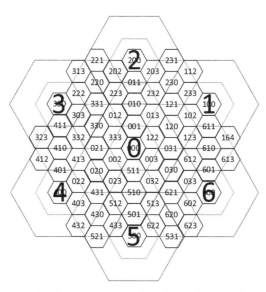

Fig. 2. The hexagonal quadtree hierarchical coding

Let the code of n level $\lambda(a) = a_1 a_2 .... a_n$ and $\lambda(b) = b_1 b_2 .... b_n$, $\oplus$ represents the coded addition operator, a_n and b_n represents the code element of each layer, the addition of codes satisfies the parallelogram rule and calculated layer by layer, shown in Eq. (1).

$$\lambda(a \oplus b) = a_1 a_2 .... a_n \oplus b_1 b_2 .... b_n = (a_1 \oplus b_1)(a_2 \oplus b_2)....(a_n \oplus b_n) \qquad (1)$$

The carry digit produced by $a_n \oplus b_n$ are abtained from Eq. (2).

$$
\begin{cases}
if\ a_i = b_i, and\ a_i, b_i \neq 0, a_i \oplus b_i = \begin{cases} a_i 0, i > 2 \\ (2a_i)0, i = 2 \end{cases} \\
if\ a_i \neq b_i, \ and\ a_i, b_i \neq 0, c = \overline{\{0, a_n, b_n\}}, \\
\qquad a_i \oplus b_i = ((a_i + b_i)/4^{|a_i - b_i| - 1}) \underbrace{c ...... c}_{n-1} \\
if\ b_i = 0, \ a_i \oplus b_i = a_i
\end{cases}
\qquad (2)
$$

For the adjacent grids searching, it can be obtained by adding it to the six adjacent codes of the center code. The six adjacent codes of the center code are $\underbrace{00...0}_{n-1}1, \underbrace{00...0}_{n-1}2,$

$\underbrace{00...0}_{n-1}3, 1\underbrace{2...22}_{n-1}, 3\underbrace{3...33}_{n-1}, 5\underbrace{1...11}_{n-1},$ as shown in Fig. 3.

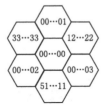

Fig. 3. The adjacent cells of hierarchical code center

3.2 Integer Coordinate Coding

The integer coordinate coding defines two coordinate axes i and j on the plane of the hexagonal grid, and use two-dimensional integer coordinate code (i, j) to describe the center of the hexagonal grid and establish indexing, as shown in Fig. 4. Based on the integer coordinate i and j, the addition and adjacent searching can be easily achieved. Figure 5 shows the adjacent cells of coordinate code.

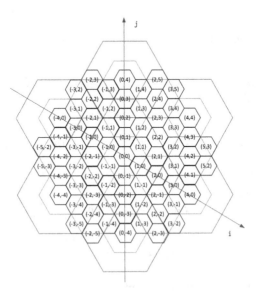

Fig. 4. The integer coordinate coding

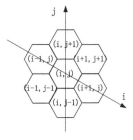

Fig. 5. The adjacent cells of coordinate code

3.3 Filling Curve Coding

A space filling curve is a one-dimensional curve that can recursively cover a specified area. This paper uses the Z-shaped space filling curve and the quaternary Morton code to implement coding. Figure 6 shows the Z space filling curve and morton code of hexagon. The coding operation of the hexagon can be achieved by Eq. (3), M_Q represents the Morton code, I_b and J_b are the row and column numbers in binary representation.

$$M_Q = 2I_b + J_b \tag{3}$$

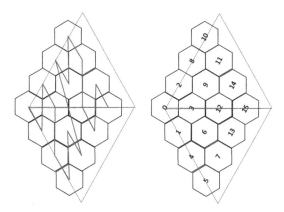

Fig. 6. The Z space filling curve and morton code of hexagon

4 Result and Discussion

In order to verify the efficiency comparison of the above algorithms, this paper implements the above three coding algorithms and calculates the computational efficiency at levels 5–10. The efficiencies of addition operation and adjacent searching are respectively the time of 100,000 random grid coding operation. The experimental results are shown in Table 1 and Fig. 7.

Table 1. Comparative result of addition and adjacent searching efficiency.

Coding level	Hierarchical coding (ms)		Integer coordinate coding (ms)		Filling curve coding (ms)	
	Addition	Adjacent	Addition	Adjacent	Addition	Adjacent
5	45.8	199.4	3.1	25.5	11.9	19.4
6	53	203.1	3.4	26.4	12.6	22
7	61.2	258	3	26.2	13.5	25.1
8	71.5	273.4	3.5	25.9	14.6	27.4
9	81.9	321.6	3.1	25.8	15.3	30
10	90.9	362.1	3.2	25.6	16.3	32.8

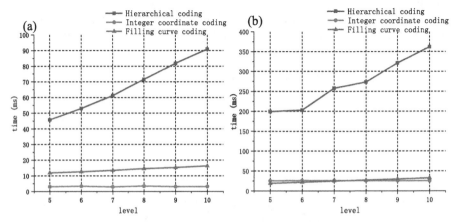

Fig. 7. The comparison of code addition efficiency (a) and adjacent searching efficiency (b)

The experiment results show that the code addition efficiency and adjacent searching efficiency of the integer coordinate coding is the best and remains unchanged with the grid level increasing. But it is described on two dimensions, which is not conducive to grid identification and storage. As grid level increases, the operation efficiency of filling curve coding is slowly decreasing. And it has certain advantages in encoding compression and storage. As for hierarchical coding, its operation efficiency is the worst, the hierarchical structure is clear, which is conductive to indexing between levels.

Generally speaking, from the perspective of coding operation efficiency and coding identification, the comprehensive characteristics of filling curve coding are better, and it is more suitable for the construction and application of discrete global grid.

References

1. Topic 21: Discrete global grid systems abstract specification [online] (2017)
2. Xuesheng, Z., et al.: Overview of the research progress in the earth tessellation grid. Acta Geod. et Cartogr. Sin. **45**(S1), 1 (2016)
3. Amiri, A.M., Samavati, F., Peterson, P.: Categorization and conversions for indexing methods of discrete global grid systems. ISPRS Int. J. Geo-Inf. **4**(1), 320–336 (2015)
4. Dutton, G.H.: A Hierarchical Coordinate System for Geoprocessing and Cartography, vol. 79. Springer, Heidelberg (1999). https://doi.org/10.1007/BFb0011617
5. Sahr, K.: Location coding on icosahedral aperture 3 hexagon discrete global grids. Comput. Environ. Urban Syst. **32**(3), 174–187 (2008)
6. Vince, A., Zheng, X.: Arithmetic and Fourier transform for the PYXIS multi-resolution digital Earth model. Int. J. Digit. Earth **2**(1), 59–79 (2009)
7. Tong, X., et al.: Efficient encoding and spatial operation scheme for aperture 4 hexagonal discrete global grid system. Int. J. Geogr. Inf. Sci. **27**(5–6), 898–921 (2013)
8. Rui, W., et al.: Encoding and operation for the planar aperture 4 hexagon grid system. Acta Geod. et Cartogr. Sin. **47**(7), 1018 (2018)
9. Wang, R., et al.: The code operation scheme for the icosahedral aperture 4 hexagon grid system. Geomat. Inf. Sci. Wuhan Univ. (2019)
10. Ben, J., Tong, X., Chen, R.: A spatial indexing method for the hexagon discrete global grid system. In: 2010 18th International Conference on Geoinformatics. IEEE (2010)
11. Mahdavi-Amiri, A., Harrison, E., Samavati, F.: Hexagonal connectivity maps for digital Earth. Int. J. Digit. Earth **8**(9), 750–769 (2015)
12. White, D.: Global grids from recursive diamond subdivisions of the surface of an octahedron or icosahedron. Environ. Monit. Assess. **64**(1), 93–103 (2000)
13. Bai, J., Zhao, X., Chen, J.: Indexing of the discrete global grid using linear quadtree. In: Proceedings of ISPRS Workshop on Service and Application of Spatial Data Infrastructure. Citeseer (2005)

Deep Transfer Learning for Successive POI Recommendation

Haining Tan[1,2(✉)], Di Yao[1,2], and Jingping Bi[1,2]

[1] Institute of Computing Technology, Chinese Academy of Sciences, Beijing, China
{tanhaining,yaodi,bjp}@ict.ac.cn
[2] University of Chinese Academy of Sciences, Beijing, China

Abstract. Personalized POI recommendation attracts more and more attention from both industrial and research fields. Due to data collection mechanism, it is common to see data collection with the unbalanced spatial distribution. For example, some cities may release check-ins for multiple years while others only release a few days of data. In this paper, we tackle the problem of successive POI recommendation for the cities with only a short period of data samples. We aim to leverage the transfer learning technique to utilize the knowledge from long-period data to enhance the POI recommendation process in target cities. Different from existing methods that transfer knowledge from one single city to a target city, we utilize the knowledge from multiple source cities to increase the stability of transfer. Specifically, our proposed model is designed as the spatio-temporal attentive recurrent neural network, MetaGRU, with a meta-learning paradigm. The spatio-temporal attentive mechanism leverage an external memory layer to store processed user historical preference information, and use spatio-temporal attention to capture the different correlations between user current status with past check-in behaviors. In addition, the meta-learning paradigm learns a well-generalized initialization of the spatio-temporal neural network, which can be effectively adapted to target cities. Extensive experiments show that our method is able to outperform state-of-the-art successive POI recommendation models in multiple tasks.

Keywords: Successive POI recommendation · Spatio-temporal attention · Meta-learning

1 Introduction

Driven by the proliferation of digital capture and the advent of the near-ubiquitous broadband internet access, location-based social networks are growing at an explosive rate. For example, Foursqure, estimated the most popular POI sharing website, stores over one billion check-ins in its repository. Therefore,

Supported by the National Natural Science Foundation of China under Grant No.: 62077044, 61702470, 62002343.

G. Pan et al. (Eds.): SpatialDI 2021, LNCS 12753, pp. 134–140, 2021.
https://doi.org/10.1007/978-3-030-85462-1_12

today's online users always face a daunting volume of candidate POIs when they search for interesting POIs from the repositories. As a result, there is an increasing demand of POI recommendation service which is able to find "interesting" or "highly related" POIs. Recommender systems provide great help to find their desired locations from a huge number of offers.

Specific to successive POI recommendation, recent years have witnessed much progress [4, 6–8], while this task is still challenging. Liu et al.[4] model the sequential influence by RNN in historical data and achieving good performance. In [7], Yin et al. propose cross-city POI recommendation model to recommend POIs when users visit another city by linking common users. More recently, Zhao et al. [8] proposed a successive POI recommendation method, which implements time gates and distance gates into LSTM to capture the spatio-temporal relation between successive check-ins for POI recommendation.

Unfortunately, user interest keeps evolving, capturing the dynamic of interest is important for interest representation regardless of the unique structure of different cities. Moreover, the user's behavior is affected discriminatively by his or her historical check-ins. What's more, the superior performance of these models, including STGCN [8], is conditioned on large-scale training data which are probably inaccessible in real-world applications. For example, there may exist only a few days of check-ins for POI recommendation in some cities. Thus, it is a more challenging task to recommend videos under the constraints of data sparsity.

Inspired by the success of the aforementioned works, we propose a spatio-temporal attentive model for personalized successive POI recommendation. The focus of our method is a computationally efficient manner to model a user's evolving interests. Our contributions are summarized as follows: (1) We investigate and formalize spatio-temporal attention based successive POI recommendation problems, increasingly important issue due to the proliferation of LBSNs and its abundant applications; (2) Designing efficient meta-learning enhanced parameter initialization during the POI recommendation procedure; (3) Through extensive experiments on real-world datasets, illustrating the efficacy and scal- ability of the designed successive POI recommendation method.

2 Methodology

2.1 Problem Statement

We first formulate the cross-city personalized POI recommendation problem. Given a set of aligned users U from data-insufficient cities, for each user $u_i \in U$, his or her check-in behaviors on LBSNs are collected. The goal of successive POI recommendation is to recommend POIs for those aligned users on LBSNS (i.e. Foursqure, Yelp et al.) through making use of the user's behaviors and mobility knowledge from other data-sufficient cities.

2.2 The Proposed Model

In this paper, we first propose a spatio-temporal attentive model(STA-GRU) using the architecture shown in Fig. 1. STA-GRU utilizes an embedding layer to capture the latent interest behind user check-ins, and leverages attentive GRU-based neural networks to model the user's sequential behavior over time, and relays the user's appetite through spatio-temporal attention weights shared across GRU. In the following, we leverage the meta-learning mechanism to transfer mobility knowledge across cities.

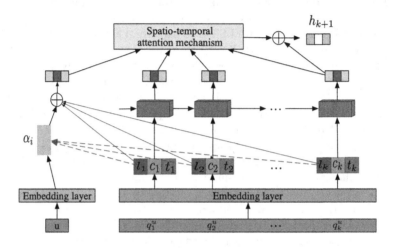

Fig. 1. The framework of spatio-temporal attentive GRU.

Spatio-Temporal Attentive Network. For the balance between efficiency and performance, we take GRU to model the latent dependency behind behaviors. The formulations of GRU are listed as follows.

$$
\begin{aligned}
\mathbf{u}_t &= \sigma \left(W^u \mathbf{q}_t + U^u \mathbf{h}_{t-1} + \mathbf{b}^u \right) \\
\mathbf{r}_t &= \sigma \left(W^r \mathbf{q}_t + U^r \mathbf{h}_{t-1} + \mathbf{b}^r \right) \\
\tilde{\mathbf{h}}_t &= \tanh \left(W^h \mathbf{i}_t + \mathbf{r}_t \circ U^h \mathbf{h}_{t-1} + \mathbf{b}^h \right) \\
\mathbf{h}_t &= (\mathbf{1} - \mathbf{u}_t) \circ \mathbf{h}_{t-1} + \mathbf{u}_t \circ \tilde{\mathbf{h}}_t
\end{aligned}
\tag{1}
$$

where σ is the sigmoid activation function,$\circ$ is the elementwise product, W and U is the transfer matrix, where $W \in \mathbf{R}^{n_H \times n_I}, U \in \mathbf{R}^{n_H \times n_H}$. n_H is the hidden size, and n_I is the input size. q_t is the input of GRU, which represents the latent embedding of the t-th behavior. In order to overcome the vanishing gradient problem, we re-weighted the output of GRUs according to the spatio-temporal attention weights, which can be seen as follows,

$$
\hat{h}_{k+1} = \sum_{i=1}^{k} \alpha_i h_i
\tag{2}
$$

where $\alpha_i = \frac{\exp(g(h_i, e_u))}{\sum_{i=1}^{k} \exp(g(h_i, e_u))}$ represents the matching score between the i-th record and the i-th latent status. $g(h_i, e_u)$ represents the attention function. We formulate the loss function of STA-GRU for each city as:

$$\mathcal{L} = \sum_{|U|} \sum_{v_+} \sum_{v_-} ln\sigma(p_{v_+} - p_{v_-}) + \mu||\phi|| \tag{3}$$

where $\sigma(x) = 1/(1 + e^{-x})$ and ϕ represents all parameters in STA-GRU.

Meta-Training Process. In meta training, we take the POI recommendation in different cities as multiple tasks. In each training batch, the model parameters are trained as follows,

$$\phi_0^s = \phi_0 - \alpha\nabla_{\phi_0}\mathcal{L}^s \tag{4}$$

Then, the model parameters can be optimized utilizing stochastic gradient descent (SGD), such that the parameters ϕ_0 are optimized as follows:

$$\phi_0 \leftarrow \phi_0 - \beta\nabla_{\phi_0} \sum_{s \in \mathcal{S}} \mathcal{L}^s |\phi_0^s \tag{5}$$

where $\mathcal{L}^s|\phi_0^s$ is the updated loss for sampled users in city s with respect to ϕ_0^s. Thus, we can obtain an initialization set of parameters ϕ_0 which can generalize well on different source cities.

Fine-Tuning on Target City for Successive POI Recommendation. When recommending POIs for target city p, MetaGRU takes Θ_0 to initialize the parameters and fine-tunes them with the historical data of p. Denoting the ground truth POI of u at $\hat{t}$ as v and y_v as the probability of related item in $\mathbf{O}$, we leverage negative sampling strategy to sample a set of negative POIs $\mathcal{V}^-$ for v. The loss function can be formulated as follows:

$$\mathcal{L}_{rec}(u) = -\sum_{v \in \mathcal{V}} \sum_{v^- \in \mathcal{V}^-} ln(y_{v^-} - y_v) \tag{6}$$

For the target city p, we first initialize Θ_p with Θ_0 and optimize as follows:

$$\Theta_p \leftarrow \Theta_p - \alpha\nabla_{\Theta_p} \sum_{u \in \mathcal{U}^p} \mathcal{L}_{rec}(u) \tag{7}$$

where $\mathcal{U}^p$ is the user set of city p. In usage, POIs with high probabilities in $\mathbf{O}$ are selected as the recommended POIs.

3   Experiments

In this section, We conduct several experiments on the available dataset collected from Foursqure[1], to show the efficiency and effectiveness of MetaGRU.

---

[1] https://foursquare.com.

Dataset. In our experiments, we use open-source check-in datasets, Foursquare, to evaluate the performance of MetaGRU. This dataset includes about 18 months (from April 2012 to September 2013) global-scale check-in data containing 33,278,683 check-ins by 266,909 users on 3,680,126 venues spanning across 415 cities.

Metrics and Baselines. We use two different metrics for performance evaluation, i.e., Hit Ration(HR@K) and Normalized Discounted Cumulative Gain(NDCG@K). For the compared baselines, we compare MetaGRU with six baselines which can be roughly divided into two groups: (1) general recommendation methods, i.e., MF-BPR [5] and CML [3]; (2) next POI recommendation methods, i.e., PRME [2], ST-RNN [4], STGCN [8], DeepMove [1];

We first evaluate the association across different methods, performances are shown in Fig. 2. Several observations can be made: the proposed MetaGRU model outperforms ST-GCN by increasing relatively 20.18% on HR@10(San Francisco), 20.53% on NDCG@10(San Francisco), 22.36% on HR@10(Los Angeles), 31.56% on NDCG@10(Los Angeles) when we only use 1-week data to train both models. Therefore, it greatly improves the performance by transferring the knowledge from multiple data sufficient cities (Table 1).

Table 1. Performance comparison of different methods.

Method	San Francisco				Los Angeles			
	HR@10		NDCG@10		HR@10		NDCG@10	
	1-week	1-month	1-week	1-month	1-week	1-month	1-week	1-month
MF-BPR	0.0383	0.0534	0.0273	0.0379	0.0401	0.0546	0.0272	0.0391
CML	0.0908	0.1261	0.0627	0.0911	0.0904	0.1251	0.0627	0.0886
PRME	0.0644	0.0889	0.0425	0.0588	0.0586	0.0826	0.0385	0.0542
ST-RNN	0.0827	0.1178	0.0526	0.0727	0.0845	0.1145	0.0525	0.0710
DeepMove	0.0937	0.1301	0.0703	0.0937	0.0968	0.1315	0.0704	0.0956
ST-GCN	0.1060	0.1466	0.0773	0.1095	0.0988	0.1409	0.0792	0.1041
MetaGRU	0.1274	0.1767	0.1056	0.1304	0.1209	0.1631	0.1042	0.1304

We further study the convergence of meta-learning procedures in terms of the number of episodes during the meta-training, as shown in Fig. 2. During the experiment, the training and testing of MetaGRU are compared in two circumstances, i.e. a single source city for initial parameter updating (SFT) and using multiple data-rich source cities to build multi-task for initial parameter updating (Multi-FT). The loss trend of the MetaGRU illustrated that both ablations of the MetaGRU can achieve the convergence of the model after a few iterations, and the meta-learning model under multiple source cities alleviates the limitations of the migration of a single city's mobile behavior knowledge to a certain extent and has a more similar generalization ability, which helps to speed up the training speed of the model and improve the recommendation effect of the model.

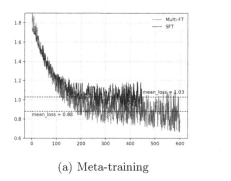

(a) Meta-training

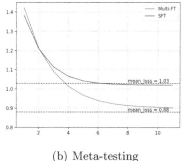

(b) Meta-testing

Fig. 2. The loss of MetaGRU against the number of epoches/update-steps in the meta-training and meta-testing in the learning task for Foursqure datasets.

4 Conclusions

In this paper, we have introduced a unified POI recommendation solution to address three typical problems in recommender systems, i.e., cold-start, model user's historical interests, vanishing gradients. We propose the spatio-temporal deep attentive model, namely MetaGRU, with the help of meta-learning. We design spatio-temporal attentive GRU to capture interest sequence particularly, which uses auxiliary loss to provide the interest state with more supervision. Furthermore, MetaGRU can effectively capture the user's evolving appetites for the POIs combining the attentive information from historical check-ins. The superior performance of our model shows that meta-learning can further boost the effectiveness of our deep structure. In the future, we will try to construct a more personalized interest model for successive POI recommendation.

References

1. Feng, J., et al.: Deepmove: predicting human mobility with attentional recurrent networks. In: WWW, pp. 1459–1468. ACM (2018)
2. Feng, S., Li, X., Zeng, Y., Cong, G., Chee, Y.M., Yuan, Q.: Personalized ranking metric embedding for next new POI recommendation. In: IJCAI, pp. 2069–2075 (2015). ijcai.org
3. Hsieh, C., Yang, L., Cui, Y., Lin, T., Belongie, S.J., Estrin, D.: Collaborative metric learning. In: WWW, pp. 193–201. ACM (2017)
4. Liu, Q., Wu, S., Wang, L., Tan, T.: Predicting the next location: a recurrent model with spatial and temporal contexts. In: AAAI, pp. 194–200. AAAI Press (2016)
5. Rendle, S., Freudenthaler, C., Gantner, Z., Schmidt-Thieme, L.: BPR: Bayesian personalized ranking from implicit feedback. arXiv preprint arXiv:1205.2618 (2012)
6. Yao, D., Zhang, C., Huang, J., Bi, J.: SERM: a recurrent model for next location prediction in semantic trajectories. In: CIKM, pp. 2411–2414. ACM (2017)

7. Yin, H., Cui, B., Zhou, X., Wang, W., Huang, Z., Sadiq, S.W.: Joint modeling of user check-in behaviors for real-time point-of-interest recommendation. ACM Trans. Inf. Syst. **35**(2), 11:1-11:44 (2016)
8. Zhao, P., et al.: Where to go next: a spatio-temporal gated network for next POI recommendation. In: AAAI, pp. 5877–5884. AAAI Press (2019)

Multiple Security Protection Algorithm for GF-2 Images Based on Commutative Encryption and Watermarking

Yu Li[1,2,3], Liming Zhang[1,2,3](✉), Hao Wang[1,2,3], and Xiaolong Wang[1,2,3]

[1] Faculty of Geomatics, Lanzhou Jiaotong University, Lanzhou 730070, China
[2] National-Local Joint Engineering Research Center of Technologies and Application for National Geographic State Monitoring, Lanzhou 730070, China
[3] Gansu Provincial Engineering Laboratory for National Geographic State Monitoring, Lanzhou 730070, China

Abstract. In the process of storage, transmission and use of GF-2 image, both data security transmission and data copyright protection should be considered. The combination of cryptography and digital watermarking is an effective way to solve this problem. However, there are some problems in the existing algorithms that combine cryptography with digital watermarking, such as high operation sequence requirements, low security, and easy leakage of keys. To solve these problems, a new multiple security protection algorithm for GF-2 image based on commutative encryption and watermarking is proposed in this paper. The integer wavelet transform is used to decompose the original GF-2 image at first, and the low frequency data is extracted and is divided into sub-block with the size of $m \times n$. Then, each sub-block is encrypted and watermarked based on the additive homomorphism of Paillier algorithm. Finally, all sub-blocks that completed the CEW operation are merged, and the merged data is performed inverse IWT to generate the encrypted-watermarked data. The proposed algorithm can achieve both security transmission by encryption and copyright protection by watermark, and the encryption algorithm and the watermarking algorithm are commutative. In addition, the watermark can be extracted from both ciphertext data and plaintext data.

Keywords: GF-2 · Commutative encryption and watermarking · Secure transmission · Copyright protection

1 Introduction

As the foundation of earth science research, GF-2 images are widely applied in many fields of earth sciences [1]. However, with the rapid development of network technology, there are series of inescapable safety issues for the high-value GF-2 images: (1) in the process of storage and transmission, data leakage is prone to occur, causing sensitive information to be illegally intercepted; (2) with the expansion of application field, data is facing serious risks such as illegal copy and dissemination, which seriously infringes on

G. Pan et al. (Eds.): SpatialDI 2021, LNCS 12753, pp. 141–147, 2021.
https://doi.org/10.1007/978-3-030-85462-1_13

the interests of data copyright owners. Therefore, it is necessary to develop an effective framework to guarantee the secure transmission and usage tracking of GF-2 images.

The combination of cryptography and digital watermarking technology provides a novel scheme to solve the above problems [2]. There are two kinds of approaches to combine cryptography with digital watermarking [3]. In the first method, watermark is embedded before encryption [4]. The disadvantage of this method is that the image must be decrypted before the watermark is extracted, which may lead to data leakage. The other method is that the watermark is embedded after encryption [5, 6]. There are two shortcomings in this method: (1) if the watermark is embedded after decryption, the encrypted image is not protected by the watermark, and the watermarked image is not protected by encryption; (2) if the watermark is embedded in the encrypted domain, the watermark can only be extracted in the ciphertext domain. According to the above analysis, combining the two kinds of technologies directly has obvious disadvantages such as high operation sequence requirements, low security, and easy leakage of keys, which is because that the encryption mechanism and the watermark mechanism affect each other.

As an emerging security technology integrating cryptography and digital watermark, commutative encryption and watermarking (CEW) provides a new idea for GF-2 image in secure storage, transmission and use tracking [7]. In work [8], stream encryption is used for CEW, in which bit planes are encrypted by XOR, and watermark is embedded in the data unaffected by encryption reversibly. In work [9], the image is divided into two datasets, the high-frequency coefficients are encrypted and the low-frequency coefficients is watermarked. However, due to the encrypted image is not protected by the watermark and the watermarked image is not protected by encryption, this method lacks security. In work [10], the homomorphism of Paillier algorithm is utilized for CEW, but the watermark is embedded in the spatial domain, so the robustness of the watermark is poor. The algorithms discussed above show that the CEW is an ideal method to combine cryptography with digital watermarking. However, GF-2 images have the characteristics of data magnanimity and stronger correlation of gray values in adjacent areas, which means that it is significant for GF-2 image watermarking technology to satisfy the invisibility and robustness of the watermark and the availability of data. Thus, above algorithms cannot be applied to GF-2 images directly.

Based on the above analysis, we present a new multiple security protection algorithm for GF-2 images based on CEW in this study, which is implemented in the identical operating domain.

2 Methodology

The purpose of the proposed algorithm is to provide the safe distribution and copyright protection for GF-2 image. The main idea is to achieve the commutativity between encryption and watermarking by utilizing the additive homomorphism of Paillier encryption algorithm. Hence, the proposed algorithm includes four compositions: key generation, watermark information generation, encryption and watermark embedding, decryption and watermark extraction. The detailed procedures are as follows.

(1) Key generation

Two prime number p and q are randomly generated based on Paillier algorithm. Their product N and least common multiple λ are calculated by Eq. (1) and Eq. (2).

$$N = p \cdot q \tag{1}$$

$$\lambda = lcm(p - 1, q - 1) \tag{2}$$

Then, the random integer g is selected to generate the public key $pk = (N, g)$ and the secret key $sk = (\lambda, \mu)$, in which $\mu = \left(L\left(g^\lambda mod N^2\right)\right)^{-1} mod N$ and $L(x) = \frac{x-1}{N}$.

(2) Watermark information generation

A binary image with copyright information is constructed as the original watermark. In order to guarantee the security of the watermark, it is indispensable to reduce the relativity between adjacent pixels. Therefore, the Arnold scrambling algorithm is used to shuffle the original watermark.

(3) Encryption and watermark embedding

Firstly, to guarantee the availability of GF-2 images, the safety of encryption and the robustness of watermark, integer wavelet transform (IWT) is performed on the original GF-2 image. The transformed low-frequency data is used as the operation area of encryption and watermark embedding, which is divided into sub-blocks of size $m \times n$.

After that, to achieve the commutativity between encryption and watermarking, the watermark embedding operation in the plaintext domain is mapped to the ciphertext domain by the additive homomorphism of Paillier algorithm.

Finally, the CEWed data is generated.

(4) Decryption and watermark extraction

The process of decryption and watermark extraction is the reverse process of encryption and watermark embedding. It is worth noting that: (1) the decryption of CEWed data needs to use the decryption key obtained from the key manager; (2) the watermark extraction of ciphertext domain is achieved by mapping the watermark extraction operation of plaintext domain to the ciphertext domain. The procedure is completed by the additive homomorphism of Paillier algorithm. The watermarked data after decryption is defined by DWed.

3   Result and Discussion

The GF-2 image of size 512×512 and the binary watermark image of size 64×64 is chosen to verify the validity and universality of the algorithm. The original GF-2 image and the binary watermark image are shown in Fig. 1.

(a) The original GF-2 image

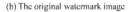
(b) The original watermark image

(c) The shuffled watermark image

Fig. 1. Experiment data

The experiments are conducted to verify the key sensitivity, relativity of adjacent pixels, watermark imperceptibility and robustness of the proposed CEW method.

(1) Key sensitivity

A secure encryption algorithm must be sensitive to the key. Hence, we evaluate the key sensitivity by modifying the decryption key. It is noteworthy that the initial key is $sk = 3960$, and the modified keys are $sk_1 = 3959$ and $sk_2 = 2043$, respectively. The decrypted images are shown in Fig. 2 using the modified keys. Obviously, the content of the original GF-2 image cannot be obtained from the DWed data with the wrong key. The original data cannot be recovered successfully even if the decrypt key is modified slightly.

(a) The decrypt result with sk=3960

(b) The decrypt result with sk1=3959

(c) The decrypt result with sk2=2043

Fig. 2. The decrypted images with the modified keys

The results of the above analysis have strong subjectivity. Hence, the mean square error (MSE) is introduced to further evaluate the key sensitivity. The experiment found that the MSE of the DWed data with the correct key is close to 0, while the MSE of the DWed data with the wrong key is extremely large. Results show that the accuracy of key must be ensured, otherwise, the original data cannot be recovered successfully even if the key is modified slightly.

(2) Relativity between adjacent pixels

The strong relativity between adjacent pixels is one of the characteristics of GF-2 images. Therefore, the relativity between adjacent pixels of the original image and the CEWed image are calculated to analyze the performance of the encryption algorithm. The relativity between adjacent pixels of the original image and the CEWed image are listed in Table 1.

Table 1. The relativity between adjacent pixels

Direction	Original image	Encrypted image
Horizontal direction	0.9814	0.0291
Vertical direction	0.9794	0.0971
Diagonal direction	0.9603	0.0207

According to Table 1, the original image has strong pixel correlation, while the pixel correlation of the CEWed image is smaller than the original image, which show that it is hard to reconstructed the original image by the adjacent pixels without the decrypt key. Therefore, the encryption security of proposed algorithm is high enough.

(3) Imperceptibility

Imperceptibility refers to the difference between the original data and the DWed data. In this section, peak signal-to-noise ratio (PSNR) is used to measure the watermark imperceptibility. In general, when the PSNR between two images is greater than 30 dB, it is difficult for the human eye to distinguish the difference between them [11], that is, the watermark is imperceptibility. According the proposed algorithm, the PSNR of the DWed data is 47.1205 dB, which shows the algorithm has good imperceptibility.

(4) Watermark robustness

It is essential for a CEW algorithm to ensure the robustness of the watermarking algorithm. To verify the robustness of the proposed algorithm, the DWed data is carried out a series of regular attacks, such as salt-and-pepper noise, gaussian noise, median filtering and cropping. The watermark robustness results of the attacked image are listed in Table 2.

According to Table 2, the watermark can be extracted after the attacked DWed data, and the NC values are greater than 0.8, which show the proposed algorithm has better robustness in resisting regular attacks, especially cropping.

Table 2. NC values of extracted watermark after image attacking

Attack type	No attack	Salt-and-pepper noise	Gaussian noise	Median filtering	Cropping	Cropping
Attack strength	0	0.01	1.0	3 × 3	1/8	1/2
NC	1.0	0.9831	0.9358	0.8754	1.0	1.0

(5) Data precision analysis

GF-2 image has higher requirement in data precision compared with the ordinary raster data. Thus, the unsupervised classification method is used to verify the precision of the DWed data. The specific classification method is K-means clustering. According to the K-Means clustering method, the DWed image is divided into five categories. The classification results of the original GF-2 image and DWed image are shown in Fig. 3. It is worth noting that there is no specific classification of terrain.

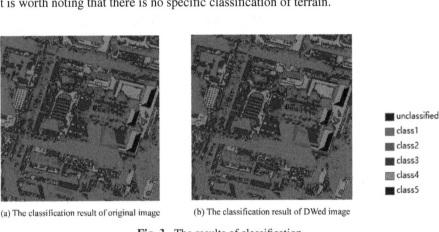

(a) The classification result of original image   (b) The classification result of DWed image

- unclassified
- class1
- class2
- class3
- class4
- class5

Fig. 3. The results of classification

It is difficult to found the classification difference between the original GF-2 image and the DWed image from Fig. 3. To further verify the data precision, the confusion matrix is used to evaluation the classification result. The result is that the overall accuracy is 99.6439% and the Kappa value is 0.9949, which shows the proposed algorithm has little effect on GF-2 image.

4   Conclusion

In this paper, a new multiple security protection algorithm for GF-2 image based on commutative encryption and watermarking is proposed. The additive homomorphism of the Paillier algorithm is used to achieve the commutativity between the encryption algorithm and the watermarking algorithm. Moreover, to guarantee the precision of GF-2 image, the integer wavelet transform is utilized to perform CEW operation. Experiments conducted confirm that the proposed CEW algorithm can extract the watermark both the cipher data and the plain data. Meanwhile, the proposed algorithm has good robustness and high security.

References

1. Zhang, X., Yan, H., Zhang, L., et al.: High-resolution remote sensing image integrity authentication method considering both global and local features. ISPRS Int. J. Geo Inf. **9**(4), 254 (2020)
2. Nasution, A.S., Wibisono, G.: An improved of joint reversible data hiding methods in encrypted remote sensing satellite images. In: Hernes, M., Wojtkiewicz, K., Szczerbicki, E. (eds.) Advances in Computational Collective Intelligence: 12th International Conference, ICCCI 2020, Da Nang, Vietnam, November 30 – December 3, 2020, Proceedings, pp. 252–263. Springer International Publishing, Cham (2020). https://doi.org/10.1007/978-3-030-63119-2_21
3. Ren, Na., Zhu, C., Tong, D., et al.: Commutative encryption and watermarking algorithm based on feature invariants for secure vector map. IEEE Access **8**, 221481–221493 (2020)
4. Puech, W., Rodrigues, J.M., Develay-Morice, J.E.: A new fast reversible method for image safe transfer. J. Real-Time Image Proc. **2**(1), 55–65 (2007)
5. Xu, Y., Xu, Z., Zhang, Y.: Content security protection for remote sensing images integrating selective content encryption and digital fingerprint. J. Appl. Remote Sens. **6**(1), 063505 (2012)
6. Ahmed, F.: Implementation of encryption and watermarking algorithm for remote sensing image. Int. J. Adv. Trends Comput. Sci. Eng. **5**(8), 17633–17637 (2016)
7. Jiang, Li., Zhengquan, Xu., Yanyan, Xu.: Commutative encryption and watermarking based on orthogonal decomposition. Multimedia Tools Appl. **70**(3), 1617–1635 (2014)
8. Zhang, X.: Commutative reversible data hiding and encryption. Secur. Commun. Netw. **6**(11), 1396–1403 (2013)
9. Benrhouma, O., Mannai, O., Hermassi, H.: Digital images watermarking and partial encryption based on DWT transformation and chaotic maps. 2015 12th International Multi-Conference on Systems, Signals & Devices (SSD), pp. 1–6. IEEE (2015)
10. Jiang, Li.: The identical operands commutative encryption and watermarking based on homomorphism. Multimedia Tools Appl. **77**(23), 30575–30594 (2018)
11. Ansari, I.A., Pant, M.: Multipurpose image watermarking in the domain of DWT based on SVD and ABC. Pattern Recogn. Lett. **94**, 228–236 (2017)

Predicting and Understanding Human Mobility Based on Social Media Check-in Data

Jing Li, Haiyan Liu$^{(\boxtimes)}$, Xiaohui Chen, Guo Wenyue, Lei Kang, Jia Li,
and Qingbo Zhao

PLA Strategic Support Force Information Engineering University, Zhengzhou 450001, Henan,
China

Abstract. Exploring the regular pattern of people's movement activities, predicting people "where to go" and "what to do" is a critical task in many applications. In this paper, we attempt to explore human mobility through social media check-in data. This study mainly includes two aspects: one is based on the considering of a multi-scale spatial characteristics of the spatiotemporal prediction method to predict the crowd volume in a certain region; one is to improve the classic TFIDF method, and propose a method to calculate the significance of POI type in the region in each period, so as to analyze and explain people's mobility tendency. We use the check-in data set of Manhattan in New York to carry out the experiment and the results show that the methods used in this paper can better predict the volume of people's activities, and realize the exploratory analysis of people's mobility tendency.

Keywords: Human mobility · Spatiotemporal prediction · Check-in data · TFIDF · Mobility tendency

1 Introduction

In recent years, with the development of intelligent sensor equipment, through GPS and other sensors, human position can be perceived by mobile devices. On social platforms, such as Foursquare, people publish text and pictures and record their location, leaving a large amount of spatiotemporal data related to life. Large scale spatiotemporal data record the moving process of people in geographical space, which hides a variety of personal preferences and human life patterns. The research on the characteristics of this mobile process has formed a common theme of epidemiology, sociology, urban planning and management, information technology, data mining and other disciplines [1].

The research of human mobility is a critical task in many applications, including navigation systems, Point-of-Interest (POI) recommendation, and location-based advertising [2], and also one of the core issues of spatiotemporal prediction [3]. It cannot only be used to analyze the traffic situation of the city, but also help to understand the functional area and population activity distribution, which has important applications in traffic management [4, 5], disaster emergency [6, 7], tourism recommendation [8, 9], urban planning [10, 11], etc.

© Springer Nature Switzerland AG 2021
G. Pan et al. (Eds.): SpatialDI 2021, LNCS 12753, pp. 148–163, 2021.
https://doi.org/10.1007/978-3-030-85462-1_14

Researchers use spatiotemporal characteristics to model and predict human movement, which can accurately predict the changes of human population in the geographical distribution with time. On the other hand, geographical location is the representation of human activities, so it cannot explain the purpose and tendency of people's activities. People's movement is always driven by the purpose of activities. Researching on human mobility from spatiotemporal data with semantics is helpful to explore the regular pattern of people's movement activities, so as to predict people "where to go" and "what to do".

In this paper, we attempt to explore human mobility through social media check-in data. This study mainly includes two aspects: one is based on the considering of a multi-scale spatial characteristics of the spatiotemporal prediction method to predict the crowd volume in a certain region; one is to improve the classic TFIDF method, and propose a method to calculate the significance of POI type in the region in each period, so as to analyze and explain people's mobility tendency. Our research has the following contributions:

- We propose a spatiotemporal prediction model considering spatial multi-scale predict crowd volume based on social check-in data. Considering the influence of spatial multi-scale on spatial feature extraction, we use different convolution kernels to extract spatial features of different spatial scales, and use parallel convolution to fuse multi-scale spatial features to obtain more detailed spatial features. What's more, spatial interaction flow is introduced into the model to extract more detailed spatial features to improve the prediction accuracy.
- We introduce the TFIDF algorithm in the field of text mining to calculate the significance of POI types, and divide the time period to compare the difference of significance in the time dimension, so as to analyze the functional tendency of human activities in geography and time.

2 Related Work

2.1 Spatiotemporal Prediction Method of Crowd Volume

The prediction of crowd volume is one of spatiotemporal prediction problem. Spatiotemporal prediction methods can be roughly divided into two categories: parametric models based on statistics and nonparametric models based on machine learning [12]. One method based on statistics is to consider spatial characteristics in traditional time prediction methods, such as considering spatial autocorrelation in autoregressive moving average model [13], and the other is to consider temporal characteristics on the basis of spatial dimension, It includes spatiotemporal geographic weighted regression model [14] and spatiotemporal prediction model based on density clustering [15–17]. This kind of method considers the spatial–temporal dependence at the same time, but it does not combine the spatial–temporal characteristics equivalently, and cannot fit the spatial–temporal nonlinear relationship well. In recent years, machine learning methods such as spatiotemporal support vector machine model [18], spatiotemporal k-nearest neighbor model [19] and deep learning method have been used in spatiotemporal prediction. However, traditional machine learning methods are constrained by the assumption that samples are independent and identically distributed, which makes it difficult to describe the non-stationarity of spatiotemporal data [20]. The deep learning method can fully mine the nonlinear characteristics of spatiotemporal data due to its efficient fitting ability to discontinuous and nonlinear data [21]. Convolutional neural networks (CNN) and graph convolutional network (GCN) can learn the spatial characteristics of spatiotemporal sequences, recurrent neural network (RNN) and its variants, long short term memory (LSTM) and gate recurrent unit (GRU) can learn the temporal characteristics of spatiotemporal series. Although the above methods can fit the spatiotemporal nonlinear relationship, most of them are based on static and single scale spatial features, and it is difficult to consider the influence of dynamic spatial interaction features and spatial multi-scale features on the learning ability and prediction results of deep learning model.

2.2 Semantic Exploration of Human Mobile Behavior Based on Spatio-Temporal Data

Social media data contains a large number of check-in locations, check-in times, and check-in place types, which can be used to study the relationship between people's check-in behavior, understand people's activity tendency, and help people establish the relationship between location and function [22]. Hu et al. [23] uses DBSCAN spatial clustering algorithm to identify the urban area of interest (AOI) from the check-in data in Flickr photos, and uses text tags to make semantic association with the identified AOI, which provides interpretability for the recognition of AOI. Based on social media data, Gao et al. [24] uses the extended Markov model to study the proportion of various human activities, and adds the time dimension to the model to study the possibility of human appearing in different positions at different times and moving between positions.

According to the literature [25], the check-in data can record the location of people in real time, and the amount of data can be accumulated over time, which is suitable for the study of people's long-term life style. Research on the spatiotemporal behavior characteristics of groups can reflect group phenomena [26]. For example, Wang et al. [27] constructs a social perception computing model based on digital footprint, and uses Sina Weibo data to explore the relationship between tourists and the behavior characteristics of tourism space. Zhou et al. [28] K-means clustering algorithm and k-nearest neighbor algorithm (k-NN) are used to identify urban people with different spatiotemporal behavior characteristics based on social media check-in data.

3   Methodology

3.1   Overview

In this paper, we attempt to explore human mobility through social media check-in data. This study mainly includes two aspects, as shown in the Fig. 1: one is based on the considering of a multi-scale spatial characteristics of the spatiotemporal prediction method to predict the crowd volume in a certain region; one is to use the improved classic TFIDF method to calculate the significance of POI type in the region in each period, so as to analyze and explain people's mobility tendency.

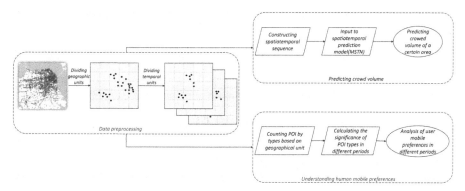

Fig. 1. Overview

3.2   Multi-scale Characteristics Spatiotemporal Network for Crowd Volume Predicting

Spatiotemporal prediction not only includes the analysis of time series, but also is affected by the spatial characteristics of entities. Based on this assumption, the multi-scale characteristics spatiotemporal network (MSTN) proposed in this paper consists of two parts: one is to extract the spatial features of different scales by local convolution neural network, the other is to capture the time features by gate recurrent unit. The main structure of the network is shown in Fig. 2. In order to better learn the spatial dependence between regions, the static flow of crowd activities in the region and the interactive crowd flow in the region are integrated into the spatiotemporal network at the same time. Time series data not only have short-term dependence, but also have long-term dependence. Through the information transmission between GRU units, the extraction and control of long-term and short-term characteristics can be realized. Finally, the learned time feature vector is input into the full connection layer, and then the activation function is used to regress to get the flow value at the next moment.

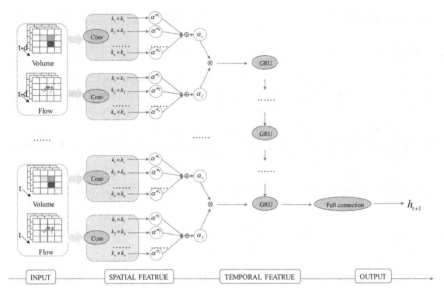

Fig. 2. Construction principle diagram of MSTN

According to the first law of geography [29], there is a certain spatial correlation of human activities. The crowd activity in a region is affected by the spatial variables of its neighborhood, and CNN can capture the local spatial features through convolution operation. On the other hand, for spatiotemporal prediction, the volume at the next moment depends on the historical volume of the region, and the flow between regions can strengthen the dynamic spatial relationship between regions. Based on this, we use CNN to extract spatial features, divides crowd volume graph and crowd interaction flow graph as network input, extracts static spatial features through crowd volume graph, and extracts dynamic spatial features through crowd interaction flow graph.

Firstly, we split the research area to n regular grids, and the whole area could be represented by region set $L = \{l_1, l_2, \cdots, l_n\}$. We split the research period to m, and the whole period could be represented by time set $T = \{t_1, t_2, \cdots, t_m\}$.

Definition 1: Crowd activity volume in the region V_l^t. We define it as the number of people passing through the grid cell in a time interval.

Definition 2: Crowd activity flow between the regions $F_{(l_i,l_j)}^t$. We define the number of people transferred from grid cell l_i to grid cell l_j in a time interval. Where, l_i is the start region of the flow and l_j is the end region of the flow.

Definition 3: Crowd activity input flow $F_{in(l_i)}^t$. It is defined as the number of people flowing into the grid cell l_i by the other grid cells in a time interval.

Definition 4: Crowd activity output flow $F_{out(l_i)}^t$. It is defined as the number of people flowing out of the grid cell l_i to the other grid cells in a time interval.

In order to extract the spatial features of the target area, local crowd volume image P_v^t and local crowd flow image P_f^t are used as network inputs for a time interval t. We set the local image size of the target area l_i to $r \times r$, and make it located in the center of the target area. The image size of P_v^t is $r \times r$, and the image size of P_f^t is $r \times r \times 2$, where

2 represents two channels, one is the local crowd input flow image, and the other is the local crowd output flow image. Therefore, the output of the local CNN is as follows:

$$\left.\begin{aligned} \alpha_v &= F\ (W_v^t P_v^t + b_v^t) \\ \alpha_f &= F\ (W_f^t P_f^t + b_f^t) \\ \alpha_r &= \alpha_v \otimes \alpha_f \end{aligned}\right\} \tag{1}$$

where "$\otimes$" tensor multiplication, α_v represents static spatial feature; α_f represents dynamic spatial feature; α_r represents spatial feature on local image scale; F represents nonlinear activation function; W and b represent the weights and biases.

Scale is one of the most important topics in Geographic Information Science [30]. Different spatial scales have certain influence on the prediction accuracy of the model. Different convolution kernels correspond to different receptive fields and can capture local features on different spatial scales [31]. As shown in Fig. 3, in this paper, the size of convolution kernel is controlled to realize the feature extraction of different spatial scales. At the same time, parallel convolution is used to flatten the multi-dimensional feature vectors extracted by convolution kernels of different sizes into one-dimensional feature vectors, and then input them into the next layer network to realize feature fusion under different spatial scales.

The output features of the fusion can be expressed as the following equation:

$$\alpha_{r(k_1,k_2,...,k_n)} = \alpha_{v(k_1,k_2,...,k_n)} \otimes \alpha_{f(k_1,k_2,...,k_n)} \tag{2}$$

where "|" represents line splicing, k represents convolution kernels of different sizes, $\alpha_{v(k_1,k_2,...,k_n)}$ represents static spatial features after fusion at different scales, $\alpha_{f(k_1,k_2,...,k_n)}$ represents dynamic spatial features after fusion at different scales, $\alpha_{r(k_1,k_2,...,k_n)}$ represents spatial features after fusion at different scales of local image scale.

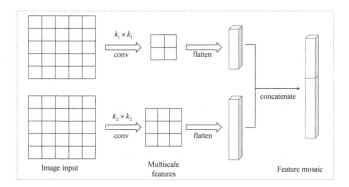

Fig. 3. Spatial multiscale feature fusion based on parallel convolution

Recurrent neural network (RNN) [32] is the most widely used neural network to process sequence data. However, the traditional RNN has the problems of gradient disappearance and gradient explosion in dealing with long-term dependence problems. As variants of RNN, LSTM [33] and GRU [34] have been proved to be able to solve the problem of long-term dependence by adding gating mechanism. LSTM and GRU have good performance in training effect. However, due to the relatively complex structure, long training time and more training parameters of LSTM, we select GRU with relatively simple structure, short training time and less parameters to capture temporal dependence. The input and output structure of GRU is shown in the Fig. 4. GRU takes the hidden state of the previous cell and the feature vector of the current time as the input, and outputs the predicted value at the next time. We formulate it as:

$$y^t = GRU(\alpha^t, h^{t-1}) \tag{3}$$

where α^t represents the spatial feature vector of time t. h^{t-1} represents the hidden state of the previous cell, which is controlled by update gate and reset gate in GRU internal structure. y^t is the prediction result.

3.3 Calculating the Time-Based Significance of Local POI Types

The POI type of check-in data can reflect the popular place type of a region. It can explain "why people like to come here", and study the types of check-in POI in different time periods can explain "why people like to come here in this period". In this paper, we calculate the significance of POI types in a region to reflect people's more preferred functional types for a region. Although the number of times of checking in POI type can reflect the type of place users like, it cannot reflect the relationship between the function and local area.

In this paper, TFIDF algorithm is used to calculate the significance of local POI types. TFIDF [35] is a statistical method used to evaluate a word for a file set. The main idea of TFIDF is: if a word or phrase appears frequently in one article and rarely in other articles, it is considered that the word or phrase has good ability to distinguish categories. We regard the local area as a document and the whole research area as a document set to calculate the significance of local POI types. At the same time, considering the influence of time factor on people's mobility tendency, we divided the time period (daytime, nighttime) to calculate the significance of POI types in different time periods. The calculation formula of the significance of local POI type based on time division is as follows:

$$S_w^t = \left(\frac{f_{l_i(w)}^t}{f_{l_i}^t}\right) \times \lg \frac{N}{n_w^t} \tag{4}$$

where S_w^t represents the significance of POI type(w) in the time period of t. $f_{l_i(w)}^t$ represents check-in frequency of POI type(w) in region l_i. $f_{l_i}^t$ represents the total number of check-in times in region l_i in time period t. N represents the number of regions divided by the whole region. n_w^t represents the number of regions where POI type(w) appears in the whole region in time period t.

4   Case Study

4.1   Data Description

We use a real data sets, New York crowd activity check-in data set (NY) to test the prediction performance of the MSTN and use the POI types' local significance to analyze the mobility preference. The data comes from foursquare platform [36]. In the experiment, 280 days of check-in data from January 1, 2012 to October 7, 2012 are selected. It totally contains 57,270 check-in records, and each record contains eight attributes as shown in Table 1. Considering the shape of the research area, the study area Manhattan in New York is divided into 20 rows × 10 columns regular grid. Considering that the sampling time of crowd check-in data is long and sparse, if the time interval is too short, the significance of the data cannot be displayed well, if the time interval is too long, it cannot reflect the periodicity of the data. Therefore, in order to avoid the sparsity of the data and consider the time semantic information, this paper divides the 24-h of the day into four time intervals, early morning (0:00–06:00), morning (06:00–12:00), afternoon (12:00–18:00) and evening (18:00–24:00). The data attributes are list in Table 1. Figure 4(a) shows the spatial distribution and grid division of the data. It can be found that there is a core check-in area, which indicates that human mobility has significant spatial characteristics. Figure 4(b) shows the number of users corresponding to different check-in times, from which we can find that there is a long tail effect between users and check-in points. Low frequency check-in users occupy the main part of the data, indicating that the check-in data set used in the experiment can reflect the typical crowd check-in behavior [37].

Table 1. Data details

Attributes	Samples
UER_ID	13,299
LATITUDE	40.782
LONGITUDE	−73.958
DATE	625
TIME	15:30
POI_TYPE	Museum
POI_TYPENU	12,348
CITY	New York

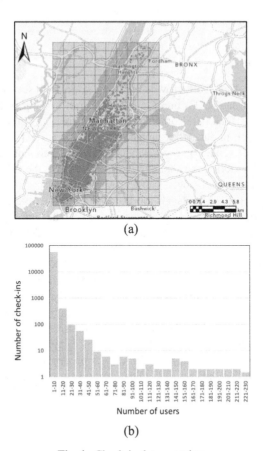

(a)

(b)

Fig. 4. Check-in data overview

4.2 Prediction of Crowd Volume

(1) Parameter setting

In order to improve the convergence speed of the model, we use the min–max normalization method to convert the data to [0,1], After the prediction, the inverse normalization of the predicted values is used for evaluation.

In the experiment, 80% of the training data is selected to learn the model, and the remaining 20% is used to verify the model. The batch size of the experiment is set to 256, the learning rate is set to 0.001. The number of GRU hidden units is an important parameter of the model, and the different number of hidden units has a great influence on the prediction accuracy. In order to select the best value, we use different hidden units to experiment, and select the best value by comparing the accuracy of the prediction. We choose the number of hidden units from [16, 32, 64, 128], testing on SFO data set, and analyze their RMSE. The result is shown in the Fig. 5. With the increase of the number of hidden cells, RMSE first show a rapid downward trend and then a gentle decline, and tend to be stable around 100. When the number of hidden units is 128, RMSE are both smallest, and the accuracy is the highest. So we set the number of hidden units to 128.

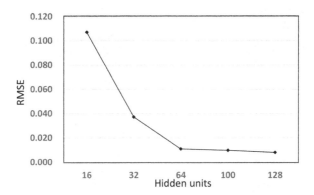

Fig. 5. Comparison of prediction accuracy under different hidden units

Initially, the number of training rounds is set to 100. However, as it shown in Fig. 6, during the training process, it is found that the loss value of the network decreases rapidly and tends to be stable at about 50 rounds. Then, the number of epochs is set to 55. During the training process, the loss values of training set and test set decreased continuously, which showed that the structure of the model was reasonable and over fitting was avoided.

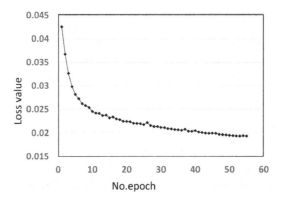

Fig. 6. Evolution of loss value with increasing number of epochs

(2) Comparison with other spatiotemporal prediction methods

The root mean squared error (RMSE) and mean absolute error (MAPE) were used to evaluate the model accuracy. The lower the value of them, the higher the accuracy of the model. In order to verify the effectiveness of this method compared with other methods, historical average (HA), autoregressive integrated moving average model (ARIMA), support vector regression (SVR), temporal graph convolutional network (T-GCN), gate recurrent unit (GRU), deep spatiotemporal residual networks (ST-ResNet) and spatial–temporal dynamic network (STDN) are selected to compare with MSTN. The results are shown in Table 2.

Table 2. Comparison with other spatio-temporal prediction methods

Method	RMSE	MAE
HA	1.2559	79.73%
ARIMA	0.5992	66.05%
SVR	1.2145	73.91%
T-GCN	0.0556	37.79%
GRU	0.0055	10.23%
ST-ResNet	0.0040	3.64%
STDN	0.0037	3.18%
MSTN	0.0026	2.92%

It can be seen that the proposed method MSTN is superior to other methods in terms of RMSE and MAPE. It can be found that:

- Traditional time series models (HA and ARIMA) and regression based methods (SVR) do not perform well in this task, mainly because these methods rely on historical data and ignore spatial factors. On the other hand, they are difficult to fit the non-stationary time series data with periodicity and trend.
- T-GCN, GRU, ST-ResNet and STDN have achieved good results in this problem, which shows that the method based on neural network can better fit the complex nonlinear spatiotemporal data. However, T-GCN performs not good as ST-ResNet and STDN, it may be that T-GCN using GCN to capture spatial features is more suitable for irregular graph structure, however, the region is divided into regular grids in this paper which is more suitable for CNN. For GRU, it only captures the time dependence and ignores the spatial characteristics. For ST-ResNet, it only captures the spatial dependence with CNN and ignores the time serise characteristics. STDN using CNN and LSTM considers the spatial features, but it only considers a single scale so that it couldn't get more abundant and prominent features. MSTN can better fit the complex nonlinear spatial–temporal relationship and take into account the spatial multi-scale features, so it has higher accuracy.

(3)   Visualization of truth and prediction

We draw the real map and forecast map of crowd activity volume in four time intervals of a day, as shown in Fig. 7. It can be found that:

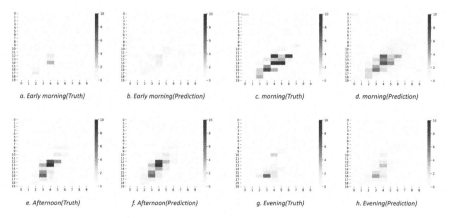

a. Early morning(Truth)   b. Early morning(Prediction)   c. morning(Truth)   d. morning(Prediction)

e. Afternoon(Truth)   f. Afternoon(Prediction)   g. Evening(Truth)   h. Evening(Prediction)

Fig. 7. Visualization of truth and prediction of crowed volume

- From the analysis of the prediction results, the prediction is close to the real value, but the prediction of the model is poor at the local minimum/maximum. This may be due to the normalization of the data. Because normalization is to enlarge and shrink the data, the data will lose part of the information about the maximum and minimum value, which may also affect the model training and prediction effect.
- From the analysis of crowd activity areas, crowd activities are mainly concentrated in fixed areas. It is speculated that there may be some well-known features in these areas, which also shows that crowd activities follow certain spatial rules;
- From the perspective of temporal semantics, the flow of the same area in different time periods is different, for example, the flow of crowd activities in the early morning is less than that in other time intervals, which may be due to the function of check-in place is caused by people's behavior habits, which also shows that crowd activities follow a certain time rule.

4.3   Analysis of Crowd Activity Preference

The POI type of check-in data can reflect the popular place type of a region. Therefore, this paper selects 253 POI types in NY dataset and divide the time period into daytime (6:00–18:00) and nighttime (18:00-the next day6:00) to calculate the time-based significance of local POI types by Eq. (4). Table 3 lists the maximum value of significance and POI type in daytime and nighttime in twenty grids. It can be seen from Table 3 that the calculated significance and the most popular POI types are different in different time periods.

Table 3. The maximum value of significance in twenty grids

GRID_ID	Max(S) (daytime)	Max(S) (nighttime)	POI_TYPE (daytime)	POI_TYPE (nighttime)
0	0.55764	0.49733	Government Building	Government Building
1	0.90982	0.97445	Island	Island
10	0.40525	0.35286	Island	Museum
12	0.42059	0.42160	Burger Joint	Burger Joint
22	0.10009	0.13032	Pier	Government Building
23	0.53900	0.32043	Harbor / Marina	Harbor / Marina
32	1.03476	0.10258	Historic Site	Historic Site
33	0.11619	0.10887	Speakeasy	Speakeasy
34	0.07045	0.06323	States & Municipalities	Cocktail Bar
35	0.13385	0.74036	Bridge	Bridge
42	0.73373	0.08720	Event Space	Event Space
43	0.31012	0.03538	Electronics Store	Electronics Store
44	0.05034	0.0416	Ramen / Noodle House	Bar
52	0.05671	0.05714	Hotel	Hotel
53	0.09947	0.04459	Park	Park
54	0.05017	0.04362	Bookstore	Music Venue
55	0.09001	0.60874	Parking	Parking
62	1.33579	0.11999	Electronics Store	Electronics Store
63	0.21194	0.03791	Sports Bar	Train Station

Figure 8 shows the significance distribution of POI types in the research area. It can be seen from Fig. 8 that the function and degree of people's activity tendency are different in different regions in the daytime and nighttime. In the daytime, the POI types in the southern region are more significant, while in the nighttime, the POI types in the central east are more significant, which indicates that there are some functional place in the south that people tend to go to during the day, and there are more suitable functional place in the central eastern regions for people's activities in the nighttime. We use tag cloud to visualize the significance of POI types in a grid in Fig. 8, from which we can see that the POI types of the grid are different during the daytime and nighttime, indicating that the functional types of attracting people in different periods of a certain area are different, such as an area is popular for office space during the daytime, but popular of dog run during the nighttime.

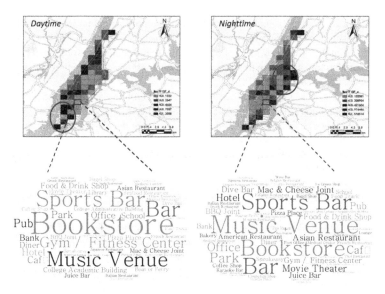

Fig. 8. Significance distribution of POI types

5   Conclusions and Future Work

In this paper, we attempt to explore human mobility through social media check-in data. This study mainly includes two aspects: one is based on the considering of a multi-scale spatial characteristics of the spatiotemporal prediction method to predict the crowd volume in a certain region; one is to improve the classic TFIDF method, and propose a method to calculate the significance of POI type in the region in each period, so as to analyze and explain people's mobility tendency. The experimental results show that the methods used in this paper can better predict the flow of people's activities, and realize the exploratory analysis of people's mobility tendency. But there are still many problems and deficiencies in the research, such as only using the regular grid method for regional division. We will try to use a more reasonable method of regional division, and try to extract different human movement patterns in the future.

References

1. Feng, L.U., Kang, L., Jie, C.: Research on human mobility in big data era. J. Geo-Inf. Sci. **16**(5), 665–672 (2014)
2. Bao, Y., Huang, Z., Li, L., et al.: A BiLSTM-CNN model for predicting users' next locations based on geotagged social media. Int. J. Geograph. Inf. Sci. **35**(4), 639–660 (2020). https://doi.org/10.1080/13658816.2020.1808896
3. Zheng, Y.: Trajectory data mining: an overview. ACM Trans. Intell. Syst. Technol. **6**(3), 29 (2015)
4. Yuan, N.J., Zheng, Y., Zhang, L., et al.: T-finder:a recommender system for finding passengers and vacant taxis. IEEE Trans. Knowl. Data Eng. **25**(10), 2390–2403 (2013)

5. Ma, X., Tao, Z., Wang, Y., et al.: Long short-term memory neural network for traffic speed prediction using remote microwave sensor data. Transp. Res. Part C **5**(4), 187–197 (2015)
6. Wang, Y., Zhou, X., Noulas, A., et al.: Predicting the spatio-temporal evolution of chronic diseases in population with human mobility data. In: International Joint Conference on Artificial Intelligence, pp. 3578–3584 (2018)
7. Balcan, D., Colizza, V., Goncalves B., et al.: Multiscale mobility networks and the large scale spreading of infectious diseases. In: APS March Meeting 2010, USA (2010)
8. Leskovec, J., Horvitz, E.: Planetary-scale views on a large instant-messaging network in the web conference, pp. 915–924 (2008)
9. Vaccari, A., Liu, L., Biderman, A., et al.: A holistic framework for the study of urban traces and the profiling of urban processes and dynamics. In: International Conference on Intelligent Transportation Systems, pp. 1–6 (2009)
10. Pan, G., Qi, G., Wu, Z., et al.: Land-use classification using taxi GPS traces. IEEE Trans. Intell. Transp. Syst. **14**(1), 113–123 (2013)
11. Wangsu, H., Yao, Z., Yang, S., Chen, S., Jin, P.J.: Discovering urban travel demands through dynamic zone correlation in location-based social networks. In: Berlingerio, M., Bonchi, F., Gärtner, T., Hurley, N., Ifrim, G. (eds.) Machine Learning and Knowledge Discovery in Databases: European Conference, ECML PKDD 2018, Dublin, Ireland, September 10–14, 2018, Proceedings, Part II, pp. 88–104. Springer International Publishing, Cham (2019). https://doi.org/10.1007/978-3-030-10928-8_6
12. Deng, M., Yang, W., Liu, Q., et al.: Heterogeneous space-time artificial neural networks for spacetime series prediction. Trans. GIS **22**(1), 183–201 (2018)
13. Jing, L., Wei, G.: A summary of traffic flow forecasting methods. Highway Transp. Res. Develop **21**(3), 82–85 (2004)
14. Huang, B., Wu, B., Barry, M.: Geographically and temporally weighted regression for modeling spatiotemporal variation in house prices. Int. J. Geogr. Inf. Sci. **24**(3), 383–401 (2010)
15. Andrienko, G., Andrienko, N., Mladenov, M., et al.: Discovering bits of place histories from people's activity trace. In: 2010 IEEE Symposium on Visual Analytics Science and Technology (VAST). IEEE, Salt Lake City, USA (2010)
16. Birant, D., Kut, A.: ST-DBSCAN: an algorithm for clustering spatial-temporal data. Data Knowl. Eng. **60**(1), 208–221 (2018)
17. Wang, S., Yao, Z., Yang, S., et al.: Discovering urban travel demands through dynamic zone correlation in location-based social networks. In: Joint European Conference on Machine Learning & Knowledge Discovery in Databases. Dublin, Ireland (2018)
18. Jiaqiu, W., Min, D., Tao, C.: Spatio-Temoral Sequence Data Analysis and Modeling. Science Press, Beijing (2012)
19. Cai, P., Wang, Y., Lu, G., et al.: A spatiotemporal correlative k-nearest neighbor model for shortterm traffic multistep forecasting. Transp. Res. Part C: Emerg. Technol. **6**(2), 21–34 (2016)
20. Shifen, C., Peng, P., Hengcai, Z., et al.: Review of interpolation, reconstruction and prediction methods for heterogeneous and sparsely distributed geospatial data. Geomat. Inf. Sci. Wuhan Univ. **45**(12), 1919–1929 (2020)
21. LeCun, Y., Bengio, Y., Hinton, G.: Deep learning. Nature **521**(7553), 436–444 (2015). https://doi.org/10.1038/nature14539
22. Yingjie, H.: Geo-text data and data-driven geospatial semantics. Geogr. Compass **12**(4) (2018)
23. Hu, Y., Gao, S., Janowicz, K., Yu, B., Li, W., Prasad, S.: Extracting and understanding urban areas of interest using geotagged photos. Comput. Environ. Urban Syst. **5**(4), 240–254 (2015)
24. Gao, L.: Research on the Temporal and Spatial Law of Human Mobility Based on Social Media. Wuhan University (2017)

25. Huang, Q., Wong, D.W.S.: Modeling and visualizing regular human mobility patterns with uncertainty: an example using twitter data. Ann. Assoc. Am. Geogr. **105**(6), 1–19 (2015)
26. Wakamiya, S., Lee, R., Sumiya, K.: Urban area characterization based on semantics of crowd activities in Twitter. In: Claramunt, C., Levashkin, S., Bertolotto, M. (eds.) GeoSpatial Semantics, pp. 108–123. Springer Berlin Heidelberg, Berlin, Heidelberg (2011). https://doi.org/10.1007/978-3-642-20630-6_7
27. Wang, L.: Spatiotemporal Behavior Analysis of Tourists Based on Social Media Geographic Data Mining. Shanghai Normal University (2017)
28. Zhou, Y., Li, Y., et al.: Analysis of classification methods and activity characteristics of urban population based on social. J. Geo-inf. Sci. **19**(09), 1238–1244 (2017)
29. Tobler, W.R.: A computer movie simulating urban growth in the detroit region. Econ. Geogr. **46**(1), 234–240 (1970)
30. Goodchild, M.F.: Models of scale and scales of modelling. In: Modelling Scale in Geographical Information Science. New York (2001)
31. Xia, C., Zhang, A., Wang, H., et al.: Modeling urban growth in a metropolitan area based on bidirectional flows, an improved gravitational field model, and partitioned cellular automata. Int. J. Geogr. Inf. Sci. **33**(5), 1–23 (2019)
32. Goodfellow, I., Bengio, Y. Courville, A.: Deep Learning. The MIT Press, Cambridge
33. Sundermeyer, M., Schluter, R., Ney, H., et al.: LSTM neural networks for language modeling. In: Conference of the International Speech Communication Association, pp.194–197 (2020)
34. Chung, J., Gulcehre, C., Cho, K., et al.: Empirical evaluation of gated recurrent neural networks on sequence modeling. Neural and Evolutionary Computing (2014)
35. Salton, G.: Term-weighting approaches in automatic text retrieval. Inf. Process. Manage **2**(4) (1988)
36. Four Square. https://www.foursquare.co.nz/
37. Bao, Y., Huang, Z., Li, L., Wang, Y., Liu, Y.: A BiLSTM-CNN model for predicting users' next locations based on geotagged social media. Int. J. Geogr. Inf. Sci. **35**(4), 639–660 (2020). https://doi.org/10.1080/13658816.2020.1808896

Research on Classification Method of Medium Resolution Remote Sensing Image Based on Machine Learning

Xiuyu Liu[1]([⊠]) [iD] and Yanyi Li[2] [iD]

[1] Country College of Geodesy and Geomatics, Shandong University of Science and Technology, Qingdao 266500, China
[2] College of Surveying and Geo-Informatics, Tongji University, Shanghai 200092, China

Abstract. Surface information extraction is an important link in geographic situation monitoring and environmental protection and plays an important role in the global sustainable development strategy. In this paper, the resolution remote sensing image in Landsat8 OLI was selected as the main data source. Aiming at the problem of the lack of classified sample data, the global land cover data in 2015 and 2017 were optimized and treated as the prior knowledge of classification. The maximum likelihood method, Support Vector Machine (SVM) and Random Forest (RF) Machine learning methods, as well as deep learning methods based on VGGNET-16 and RESNET-18 models, were used to compare and study surface information extraction methods in the Yellow River Delta region. The results show that the above method is highly feasible. Based on the feature optimization, the overall classification accuracy of RF and SVM models in the machine learning method is high, and the classification accuracy of RF and SVM models is up to 87.3% and 86%. In the deep learning algorithm, the classification accuracy of VGGNET-16 and RESNET-18 models is greatly improved compared with the machine learning method. The classification accuracy of RESNET-18 is up to 94.1%, and the Kappa coefficient is 0.91. The research method in this paper has good applicability and popularization value in the classification of medium resolution remote sensing objects.

Keywords: Remote sensing classification · Surface information extraction · Machine learning · Depth learning · Landsat8 OLI

1 Background of the Problem

Land use and land cover change have an important impact on the analysis of global energy balance and terrestrial ecosystem diversity [1]. The traditional manual interpretation methods can not meet the actual production needs in terms of efficiency and accuracy. How to realize the automatic and intelligent extraction of LULC information has become one of the current research hotspots. With the progress of modern machine learning and deep learning technology, this technology is also developing. As far as the

G. Pan et al. (Eds.): SpatialDI 2021, LNCS 12753, pp. 164–173, 2021.
https://doi.org/10.1007/978-3-030-85462-1_15

research field of this paper is concerned, from the beginning, the classification extraction method is based on artificial feature construction, and the extraction method is based on machine learning. For example, from the maximum likelihood method based on probability model, or unsupervised classification model based on pixel similarity, to the decision tree classification algorithm based on prior knowledge, and the emerging extraction method based on deep learning [2]. Nowadays, the extraction method of deep learning can automatically extract the classification information of land cover layer by layer to classify remote sensing images and has achieved good results at present [3]. At the same time, with the development of computer technology and machine learning technology, SVM algorithm [4] based on image feature difference and random forest algorithm [5] suitable for small sample classification proposed by Yang Junwen and others are gradually formed, which have achieved good results in the case of the small sample size of remote sensing data. With the rapid development of deep learning in the field of artificial intelligence, remote sensing image abstract features are extracted automatically based on the deep learning method. Compared with the traditional method of artificial feature construction, the former can effectively express the essential features of data. Karalas et al. Applied deep learning to multi-label land cover classification task [6] and creatively applied semantic segmentation method to remote sensing land surface classification, which greatly improved the automation and level of LULC information extraction. However, the vast majority of researchers are based on high-precision prior knowledge for training and feature extraction. Prior knowledge is often deep learning data set, which has the limitations of the small amounts of data, few types of data, low sharing, and high cost, so it is difficult to be widely used and can not meet the needs of the current massive remote sensing data analysis and utilization. Therefore, how to obtain a certain precision, a large number of reliable and low-cost prior knowledge, and extract high-quality training samples for training, to obtain high-precision classification results, has become the main problem restricting the intelligent extraction of surface information. Therefore, samples and their features are the keys to realize remote sensing image classification. With the continuous development of earth observation and land surveys in China, abundant LULC historical data have been accumulated [7], which can provide reliable prior knowledge for land classification research. Because the time of obtaining LULC information has a certain span, and there are some technical differences, these data have some problems, such as the poor current situation, the inconsistency between the land type map and the actual land cover type, and most of them are local accurate prior knowledge [8].

To solve these problems, this paper proposes a remote sensing image classification method based on local accurate prior knowledge. To solve the problem of local surface feature classification errors in the surface coverage classification data set, firstly, the surface type information from different sources and different periods is fused and optimized to avoid abnormal samples, to obtain more ideal and reliable prior knowledge. Finally, based on particle swarm optimization genetic algorithm (PSO-GA) for feature extraction and selection, machine learning and deep learning algorithm are used to realize the extraction of LULC information from remote sensing images.

2 Research Basis

2.1 Overview of the Study Area

The Yellow River Delta is located in the West Bank of Laizhou Bay and the South Bank of Bohai Bay. Due to the short time of land formation, the impact of land-sea interaction, and human production activities, the ecological environment of the delta is relatively fragile, and the surface information is complex [9–11]. Therefore, it is of great significance to select the Yellow River Delta as the research area.

2.2 Introduction of Data Sources

The main data used in this paper are Landsat-8 oli image data. Based on the data of global land cover classification 2015 [12], global land cover classification 2017 [13], and the 30 m resolution spatial distribution data set of coastal aquaculture ponds in China (2015) [14], the sample data set is established as a priori knowledge, and it is used as the basic data for classification method research. The remote sensing image is fused by the gram Schmidt fusion algorithm, which can better preserve the spectral characteristics of the image. The spatial resolution of the fused oli remote sensing image is 15 m.

2.3 Sample Data Preprocessing

Due to the difference of acquisition time between remote sensing image and surface coverage data, the actual features and surface information are not completely consistent, so the existing surface classification data can not be regarded as completely correct. By comparing with the actual image features, combined with Google Earth historical image data, it is found that there are some defects in the classification data of the study area in 2015 and 2017, in which the 17-year classification data has higher classification accuracy for villages and towns, and the 15-year classification data has higher classification accuracy for urban areas and bare land. The accuracy of the 30 m resolution spatial distribution data set of Chinese coastal aquaculture ponds is very high, which is consistent with the actual features.

In this study, the above classification data sets are fused and merged:

(1) The land cover data in 2015 and 2017 were merged.
(2) The bare land category in 2015 land cover data is reserved.
(3) The data of the previous step and the spatial distribution data of coastal aquaculture ponds were superposed and analyzed.
(4) The obtained data were recorded, and the forest, grassland, shrub, tundra, and bare land were classified into other categories.
(5) Based on the probability model statistics of the obtained results, the probability distribution range of the pixel eigenvalues of each surface coverage type is analyzed, the normal threshold range is determined, the outlier data is marked, and the prior knowledge of "assumption error" is removed.

On the whole, the change and classification error of the plot is small. Through the above methods, we can identify most of the error prior knowledge, avoid the abnormal samples, and get the classification samples with relatively high accuracy.

2.4 Research Technology Route

Firstly, the original remote sensing image and sample data are preprocessed, including radiometric calibration and atmospheric correction; The sample data are fused, merged, and the abnormal samples are identified. Then, PSO-GA algorithm is used to extract and select features, and different feature combinations after fusion are normalized to facilitate model training. The machine learning methods such as SVM and RF and the deep learning methods of VggNet-16 and ResNet-18 are used to extract LULC information, and compared with the traditional maximum likelihood classification methods, the characteristics and advantages of different algorithms in the surface information extraction of medium resolution remote sensing images are analyzed.

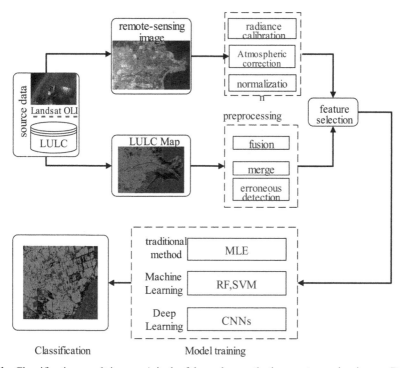

Fig. 1. Classification result image; A is the false color synthetic remote sensing image, B is the label, C is the RESNET-18 model, D is the random forest algorithm.

3 Experimental Process

3.1 The Construction of Land Classification System

There are six types of LULC information in the study area, including water body, cultivated land, impervious surface, beach, coastal aquaculture and others.

3.2　Research Technology Route

In order to better highlight the differences between different features, the image features extracted in this paper include spectral features, texture features and index features, a total of 38 feature parameters. The first principal component of landsat-8 oli multispectral is obtained by PCA transform, and the texture feature is extracted by GLCM. Using the feature combination of random forest algorithm when the difference of cross-validation accuracy before and after is less than 3% as the initial set, the selection, crossover, and mutation operator steps of genetic algorithm are added to particle swarm optimization algorithm, which not only retains the advantages of fast convergence of PSO but also increases the advantages of GA operator. At the same time, it improves the diversity of the population and the global search ability to prevent the algorithm from falling into the local optimal solution [15, 16]. Based on many experiments, the most suitable particles are selected as the optimal feature set, which includes 13 feature categories, including 5 spectral features, 6 exponential features and 2 texture features.

The specific feature categories are shown in the table below (Table 1).

Table 1. Feature categories

Feature information	Feature name
Spectral characteristics	Coastal band, blue band, red band, green band, near infrared band, short wave infrared band 1–2
Index characteristics	NDVI MNDWI EVI PCA1 PCA2 PCA3
Texture features	Mean, variance, synergy, correlation

3.3　Research Technology Route

In this paper, feature combination is carried out on the basis of feature selection of PSO-GA algorithm. A total of six experimental schemes are designed, as shown in Table 2. All the features to be selected in each scheme are internally normalized to eliminate the influence of different features due to different values and types. Through the combination of different bands, the characteristics and advantages of various classification methods are studied.

Table 2. Experimental scheme design

Programme	Example
I	Blue band, red band, green band
II	Blue band, red band, green band, near infrared band, short wave infrared band 1
III	Blue band, red band, green band, near infrared band, short wave infrared band 1, NDVI, MNDWI
IV	Blue band, red band, green band, near infrared band, short wave infrared band 1, NDVI, MNDWI, NDBI, mean, variance
V	PCA1, PCA2, PCA3, NDVI, MNDWI
VI	PCA1,PCA2,PCA3,NDVI,MNDWI, NDBI, Mean, Variance

3.4 Classification Experiment of Machine Learning and Deep Learning

The machine learning method uses support vector machine algorithm and random forest algorithm for image classification and uses the traditional maximum likelihood algorithm (MLE) for accuracy comparison. Based on the optimization of PSO-GA algorithm, a random forest algorithm is used to select the parameter combination with the minimum OOB error. In this study, the sample data is divided into two parts, 70% as training data and 30% as validation sample data.

When initializing the machine learning model, the parameters of MLE algorithm are set as follows: the likelihood threshold of the model is not set, the radial basis kernel function is used as the basic kernel function type, and the reciprocal of the number of bands is taken as the gamma value. At the same time, considering the background of the basic problem in this paper, we also need to set the model parameters of SVM and RF algorithm. Among them, the parameters of SVM algorithm are set as follows: the penalty coefficient is set to 100, the classification level is set to 0, and the classification probability threshold is also set to 0. The parameters of RF algorithm are set as follows: the number of random forest trees is set to 100, and Gini coefficient is selected as impurity function. After data processing, the effect of the machine learning method is shown in the table below (Table 3).

Table 3. Experimental scheme design

Method Programme	MLE	RF	SVM
II	76.24%	84.89%	85.13%
III	73.76%	85.27%	84.82%
IV	73.66%	86.71%	83.28%
V	79.25%	86.95%	85.15%
VI	79.34%	87.26%	86.04%

The accuracy of maximum likelihood method is 73%–79%, and that of random forest and support vector machine is 8%–13% higher than that of maximum likelihood method. In the random forest method, the overall classification accuracy of scheme IV is about 2% higher than that of scheme II, in which the classification accuracy of cultivated land is 7%, and the classification accuracy of impervious surface and tidal flat is more than 90%. Due to the large distribution of wetlands in the Yellow River Delta, the error of water, cultivated land and mariculture is large. After adding NDVI, MNDWI and texture features to scheme IV and VI respectively, the error of water classification is greatly reduced. In scheme II, the accuracy of RF and SVM is almost the same. With the increase of data dimension, the accuracy of RF classification is about 3% higher than that of SVM classification, but the time-consuming of SVM algorithm is 5–7 times that of RF algorithm.

The deep learning method uses VggNet-16 deep convolution neural network and ResNet-18 deep convolution neural network to extract LULC information. Compared with machine learning algorithm, deep learning algorithm can better extract the deep information of the image. Machine learning is a combination of linear transformation and nonlinear transformation in the process of input and output, while deep learning can fully learn the complex information of image [17]. In order to make the training samples more diverse and avoid overfitting phenomenon, which leads to the lack of generalization ability of the model, the data is expanded by using image flipping, rotation, adding Gaussian noise points and other operations, and 5 is selected for the classification pixel centered 5×5 neighborhood is upsampled by bilinear interpolation to 244×244, and then trained the neural network. The model is built based on tensorflow2.0, and the parameters are shown in Table 6. Scheme I and VI are selected for classification. Overall accuracy (OA) and kappa coefficient are used as the evaluation indexes.

After processing, it is found that the accuracy of resnet-18 in scheme I is 92.8%, 1.5% higher than that of vggnet-16, and the kappa coefficient is 0.88, 0.06 higher than that of vggnet-16. With the increase of the number of features in scheme VI, the accuracy of the two models is 1.1%–1.3% higher than that in scheme I, and the kappa coefficient is also improved. Compared with the machine learning model, the accuracy of resnet-18 model is improved by 6.84% and 8.06% compared with random forest algorithm and support vector machine, and by 14.7% compared with the traditional maximum likelihood method. The classification accuracy of water body and tidal flat is more than 96%, and impervious surface is more than 90%. But the overall time-consuming is 4–7 times more than machine learning method, and the model parameters are more complex.

4 Experiment and Result

In this paper, machine learning and deep learning are used to extract LULC information from remote sensing images based on local accurate prior samples. Based on different feature fusions, different methods have their characteristics. The classification results based on scheme VI are shown in Figs. 1 and 2 (Fig. 3).

The main conclusions are as follows:

(1) In the machine learning method, the accuracy of the random forest and support vector machine model is 10%–13% higher than that of the traditional maximum

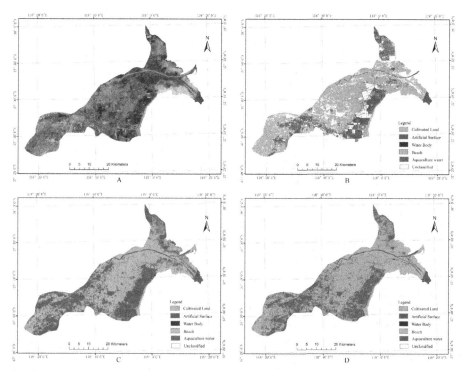

Fig. 2. Classification result image; A is the false color synthetic remote sensing image, B is the label, C is the RESNET-18 model, D is the random forest algorithm.

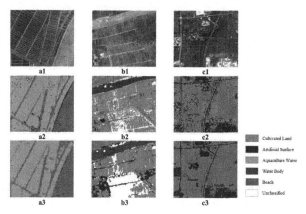

Fig. 3. Result of Classification; A1, B1, and C1 are the original images. A2, B2, and C2 are the results of random forest classification of Scheme VI, and A3, B3, and C3 are the results of the RESNET-18 model of Scheme VI.

likelihood method. In scheme II, the classification accuracy of RF and SVM is almost equal to 85%. With the optimization of features and the increase of data dimension, the accuracy of RF algorithm is improved by about 3%, and the classification accuracy of SVM is reduced by 2%. The surface information of the study area is complex, and the error of water and cultivated land, mariculture is large. Compared with schemes IV, VI and II, the error of water and cultivated land classification is greatly reduced, and the accuracy of cultivated land classification is improved by about 10%. And the overall time consumption is 5–7 times of RF algorithm. It can be concluded that RF algorithm model has strong stability, good anti-noise, processing high-dimensional image data has certain advantages, high computational efficiency, fewer parameters, and good generalization ability. When the dimension of SVM is not too high, it can achieve the same accuracy as RF algorithm, but it takes a long time. MLE algorithm has the advantages of simple operation, shortest processing time, and low precision.

(2) In the deep learning algorithm, the scheme I only contain the red, green and blue band data of remote sensing image, and the classification results can reach 91.3%–92.8%. The accuracy of ResNet-18 is 1.5% higher than that of VggNet-16, and the kappa coefficient is 0.88, which is 0.06 higher than that of VggNet-16. In scheme VI, with the increase of feature dimension, the accuracy is improved by 1.1–1.3%, and the highest classification accuracy is 94.1%. Compared with RF algorithm and SVM, the accuracy is improved by 6.84% and 8.06%, and compared with the traditional maximum likelihood method, the accuracy is improved by 14.7%. The classification accuracy of water body and beach is more than 96%, and impervious surface is more than 90%. However, the training time of the network model is 5–8 times more than that of the machine learning method, and the adjustment complexity of the model parameters is high.

(3) In Figs. 1 and 2, it can be seen that there are many noise points in the classification results of the machine learning method. The deep learning algorithm can extract the classification results of large area and serial pieces. The error is low, the accuracy is high, and the actual ground objects are in high agreement. The results show that CNN neural network can extract the deep features of the image more comprehensively by combining the convolution layer and pool layer, so as to learn better, and has a better adaptability and high classification accuracy.

5   Conclusions

In order to solve the problem of few classification sample data, this paper takes the Yellow River Delta as the research area and studies the application of medium resolution remote sensing image classification based on local accurate prior knowledge. Based on fusion and merging of land cover information in different periods, removing "false assumption" samples and other preprocessing, PSO-GA algorithm is used for feature extraction and selection. The accuracy of RF model optimized by PSO-GA algorithm is 87.3%, and ResNet-18 algorithm is 94.1%. The results show that both machine learning and deep learning methods can achieve better classification accuracy. Compared with machine learning algorithm, the deep learning algorithm can extract the deep information of images better. Machine learning is a combination of linear transformation and nonlinear transformation in the process of input and output, while deep learning can fully learn the complex information in the image, so the latter has higher accuracy than the former. The research method in this paper has good applicability and promotion value in the classification of medium resolution remote sensing features.

References

1. Sulla-Menashe, D., Gray, J.M., Parker Abercrombie, S., Friedl, M.A.: Hierarchical mapping of annual global land cover 2001 to present: the MODIS Collection 6 Land Cover product. Remote Sens. Environ. **222**, 183–194 (2019). https://doi.org/10.1016/j.rse.2018.12.013
2. Zhu, L.A.N.: Research on Artificial Surface Extraction Method Based on Medium and High Resolution Remote Sensing Image. University of Electronic Science and technology, Chengdu (2020)
3. Shuang, Li., Erxun, Z.: Research on remote sensing image classification based on decision tree. Reg. Res. Dev. **01**, 17–21 (2003)
4. Anthony, G., Gregg, H., Tshilidzi, M.: Image classification using SVMs: one-against-one vs one-against-all [EB/OL] (2017). https://arxiv.org/abs/0711.2914
5. Yang, J., Zhang, J., Zhu, X., Xie, D., Yuan, Z.: Application of random forest in hyperspectral remote sensing data dimension reduction and classification. J. Beijing Normal Univ. (NATURAL SCIENCE EDITION) **51**(S1), 82–88 (2015)
6. Konstantinos, K., Grigorios, T., Michalis, Z., Panagiotis, T.: Deep learning for multi-label land cover classification. SPIE Remote Sensing (2015)
7. Xu, X., Du, Z., Zhang, H., Feng, L., Shen, X.: Driving forces of land use/cover change in northern Shanxi from 1986 to 2010. China Environ. Sci. **36**(07), 2154–2161 (2016)
8. Xu, C., Chen, Z.H.: Landsat 8 oli image deep learning classification method integrating inaccurate prior knowledge. Comput. Appl. **40**(12), 3550–3557 (2020)
9. Yan, M., Gengxing, Z.: Evaluation of ecological environment in estuary area based on satellite remote sensing data: a case study of Kenli County in Yellow River Delta. China Environ. Sci. **29**(02), 163–167 (2009)
10. Liu, G., Liu, Q., Ye, Q., Chang, J.: Dynamic monitoring of land use and integrated coastal zone management in the Yellow River Delta. Resour. Sci. (05), 171–175 (2006)
11. Yang, H., Li, J., Li, X., Huang, S., Xu, M.: Analysis of land use and ecological environment landscape in the estuary area of the Yellow River Delta. Water Conserv. Hydropower Technol. **03**, 82–84 (2007)
12. Ji, L., Gong, P., Geng, X., Zhao, Y.: Improving the accuracy of the water surface cover type in the 30 m FROM-GLC product. Remote Sens. **7**(10), 13507–13527 (2015). https://doi.org/10.3390/rs71013507
13. Li, C., et al.: The first all-season sample set for mapping global land cover with Landsat-8 data. Sci. Bull. **62**(07), 508–515 (2017)
14. Ren, C., et al.: Rapid expansion of coastal aquaculture ponds in China from Landsat observations during 1984–2016. Int. J. Appl. Earth Observ. Geoinf. **82**, 101902 (2019)
15. Li, H., Peng, Y., Deng, C., Gong, D.: Review on hybrid research of GA and PSO. Comput. Eng. Appl. **54**(02), 20–28+39 (2018)
16. Shi, D.: Remote sensing image classification based on GA-PSO optimized hierarchical dt-svm hybrid kernel and its application [D]. Nanjing University of Posts and telecommunications (2014)
17. Krizhevsky, A., Sutskever, I., Hinton, G.E.: ImageNet classification with deep convolutional neural networks. **60**(6), 84–90 (2017)

Wavelet Threshold Denoising and Pseudo-range Difference-Based Weighting for Indoor BLE Positioning

Yang Dai[✉] and Jianqiang Wang

Faculty of Geomatics, East China University of Technology, Nanchang 330013, China

Abstract. The ranging algorithm of Received Signal Strength Indication (RSSI) is one of the most frequently used algorithms for Bluetooth Low Energy (BLE) indoor positioning. In order to solve the problems of the insufficient RSSI ranging accuracy and low filtering accuracy, an algorithm of wavelet threshold-based denoising and pseudo-range difference weighted is proposed. This method uses the algorithm of wavelet threshold denoising to eliminate the environmental noise and the algorithm of rejecting abnormal signals to improving the Kalman filtering accuracy. Furthermore, the traditional weighting method by the inverse of physical distance is not accurate enough because of the insufficient physical distance caused by the channel transmission model (CTM) parameter error. In view of the above problems, a new weighted algorithm is studied by using the pseudo distance differential method. Finally, Gauss-Newton iterative algorithm is used to calculate the coordinates of indoor target points. Experiment results demonstrate that the average positioning error of the proposed algorithm is about 2 m, which is significantly better than the positioning accuracy of the mean filtering algorithm, gauss filtering algorithm, Kalman filtering algorithm, and wavelet Kalman filtering algorithm.

Keywords: Indoor positioning · Kalman filtering · Wavelet threshold denoising · Gauss-Newton iterative · Differential weighting

1 Introduce

With the development of BLE 5.0 technology and the popularity of various intelligent devices, wireless sensor network (WSN) assisted indoor positioning has become a research hot-spot of Location Based Services (LBS) [1]. Based on Received Signal Strength Indication (RSSI), BLE positioning technology (BLE-RSSI) [2] is a classic method for WSN node positioning. This positioning technique is generally divided into two parts: the RSSI ranging model and the positioning algorithm. The former measures the distance between an Access Point (AP) and a mobile terminal based on the CTM by gathering RSSI. The latter utilizes trilateration positioning, centroid algorithm or fingerprint algorithm [3] to determine the target point coordinates. In fact, there are wireless signal attenuation and multi-path interference [4] in a complex indoor space. Therefore,

© Springer Nature Switzerland AG 2021
G. Pan et al. (Eds.): SpatialDI 2021, LNCS 12753, pp. 174–185, 2021.
https://doi.org/10.1007/978-3-030-85462-1_16

how to correctly estimate RSSI is an urgent issue to improve indoor wireless positioning accuracy.

In RSSI positioning, there are three high frequently used ways which are the mean filtering (MF), Gaussian filtering (GF) [5] and Kalman filtering (KF) [6]. Xue et al. [7] utilized mean filtering on RSSI observations and the weight coefficient is determined by physical distance which is converted from the CTM. In this way, the positioning accuracy is about 4 m by K-nearest neighbor (KNN) algorithm. This way ignores some problems such as insufficient RSSI ranging accuracy caused by the environment noise. Zhou et al. [8] utilized Kalman smooth RSSI and traditional Least Squares (LS) positioning. This method does not need a lot of time to build a fingerprint database, however the results of the Kalman filter are affected by the initial value. The above filtering methods cannot distinguish the real signal from the noise, and are not obvious for improving the positioning accuracy. Sari R et al. [9] utilized the Maximum Likelihood (ML) nonlinear equations to estimate distance. The positioning is then carried out using the classical LS method. This method leaves out of consideration the influence of abnormal signals on the ML estimation results.

In response to the problems mentioned above, we proposed a new BLE-RSSI positioning algorithm based on wavelet threshold denoising and pseudo-range difference weighting in this paper. Wavelet threshold denoising and Kalman filtering are first used to preprocess RSSI observation signals. A set of stable BLE wireless signal is obtained by eliminating environmental noise and deleting abnormal signals. Fitting by linear regression algorithm [10], the CTM parameters are then carried out to improve the RSSI ranging accuracy. A pseudo-range differential weighting method is proposed in order to reduce the weight coefficient error. Compared with the traditional weighting method, this one only needs one parameter in the CTM. Finally, the coordinates of indoor target points are obtained by Gauss Newton iterative algorithm [11]. Experiment results demonstrate that the positioning accuracy of the proposed algorithm is significantly better than other methods, which are mean filtering, Gaussian filtering, and Kalman filtering.

2 Channel Transmission Model

2.1 RSSI Ranging Principle

Based on BLE-RSSI, the principle of the distance measurement model is that: the intensity of the electromagnetic signal from the indoor iBeacon node is obtained through the receiver, and then the CTM is used to convert it into distance. It can be described as [9]

$$Pr(d_i)_{dBm} = Pr(d_0)_{dBm} + 10\gamma \lg\left(\frac{d_i}{d_0}\right) + \xi_{dBm} \tag{1}$$

where $Pr(d_i)_{dBm}$ and $Pr(d_0)_{dBm}$ are the signal strength level of the receiver when the distance from the beacon node is d_i and d_0. γ is the path attenuation index [9]. ξ_{dBm} represents the environmental noise following the Gaussian distribution of expectation 0 and variance σ^2. In the experiment: we get $\xi_{dBm} = 0_{dBm}$, $d_0 = 1$ m to represent special testing area. $Pr(d_0)_{dBm}$ is the received power at $d_0 = 1$ m. So the distance between the receiver and the beacon node can be transformed into

$$d_i = 10^{\left(\frac{Pr(d_i)_{dBm} - Pr(d_0)_{dBm}}{10\gamma}\right)} \tag{2}$$

2.2 BLE Digital Signal Processing

In order to improve the RSSI ranging accuracy, the RSSI observations is processed by wavelet threshold denoising [12]. The abnormal signal is removed by Gaussian criterion. After that, the digital signals without noise is obtained. Finally, one group stable signal is obtained through Kalman filtering.

The digital signal $RSSI(i)$, which usually contains noise, it can be expressed as

$$RSSI(i) = rssi(i) + \beta \cdot e(i), (i = 1, 2, \cdots n) \tag{3}$$

where $rssi(i)$ is the real signal, $e(i)$ is the ambient noise, and β is the noise intensity.

The original signal with length n is decomposed into three layers and different wavelet coefficients are generated. These coefficients are preserved whose amplitude is higher than the global unified threshold $\beta(2log(n))^{1/2}$. Then the processed wavelet coefficients are reconstructed by inverse wavelet transform, and finally the stable wireless signal is recovered.

$$\begin{aligned} RSSI &= CA_1 + CD_1 = CA_2 + CD_2 + CD_1 \\ &= CA_3 + CD_3 + CD_2 + CD_1 \end{aligned} \ (i = 1, 2, 3) \tag{4}$$

In formula (4), CA_i is the approximate part of the decomposition, and CD_i are the detailed part.

After wavelet threshold denoising, $rssi$ approximates to (μ, σ^2) Gaussian distribution, and its probability density function is

$$f(rssi_i) = \frac{1}{\sigma\sqrt{2\pi}} \exp\left[-\frac{(rssi_i - \mu)^2}{2\sigma^2}\right] \tag{5}$$

where μ and σ^2 separately represents expectation and variance of $rssi$.

$$\mu = \frac{1}{n}\sum_{i=1}^{n} rssi_i, \sigma^2 = \frac{1}{n-1}\sum_{i=1}^{n}(rssi_i - \mu)^2$$

The $rssi$ falling outside the interval $(\mu - 3\sigma, \mu + 3\sigma)$ is deleted as an abnormal signal [5]. The average value of $rssi$ signal in this interval is taken as the initial value of Kalman filtering. The system state equation and observation equation are obtained as

$$\begin{cases} rssi(k) = A \cdot rssi(k-1) + W(k-1) \\ Z(k) = H \cdot rssi(k) + V(k) \end{cases} \tag{6}$$

As a recursive prediction algorithm, the Kalman gain $K(k)$ is obtained using predicted value $rssi(k)$ and the predicted covariance $P(k)$ in the Kalman filtering [13]. Finally, the indoor wireless signal state and covariance are then updated by Kalman gain.

$$rssi(k|k-1) = A \cdot rssi(k-1|k-1) \tag{7}$$

$$P(k|k-1) = AP(k-1|k-1)A^T + Q \tag{8}$$

$$K(k) = HP(k|k-1)H^T \cdot \left(HP(k|k-1)H^T + R\right)^{-1} \tag{9}$$

$$rssi(k|k) = rssi(k|k-1) + K(k)(Z(k) - Hrssi(k|k-1)) \tag{10}$$

$$P(k|k) = (I - K(k)H)P(k|k-1) \tag{11}$$

In formulas (6) to (11), $rssi(k)$, $z(k)$ are the system state values and measured values at time k. A is the state transition matrix, H is the observation matrix, and I is the identity matrix. $W(k)$, $V(k)$ are the environment noise following the $(0, Q)$ and $(0, R)$ distribution. Finally, the mean algorithm is used to deal with the filtered digital signal and bring it into formula (2) to calculate the physical distance.

2.3 CTM Parameter Estimation

According to formula (2), the RSSI ranging accuracy depends on the accuracy of indoor CTM parameter γ and $Pr(d_0)_{dBm}$. In traditional positioning method, random sets of CTM parameters will result in the lack of ranging accuracy and affect the final positioning accuracy. Therefore, it is necessary to explore a more suitable parameter estimation model for the current environment. The RSSI preprocessing method is used to deal with the RSSI observation values of specific beacon nodes at different distances. Then the parameters γ and $Pr(d_0)_{dBm}$ are fitted through the linear regression algorithm. By removing the RSSI with a large deviation, the last fitting parameters $Pr(d_0)_{dBm}$ and γ can be output when the average error between the remaining RSSI and the fitted curve is less than 0.5 dBm.

3   Positioning Algorithm

3.1   Least Squares Positioning Algorithm

Let $AP_i(x_i,y_i)$, $(i = 1,2,...,M)$, $X(x,y)$, $RP(x_{rp},y_{rp})$ be the coordinates of M beacon nodes, N target points and one reference point (RP) in the independent coordinate system of the closed room. Assuming d_i be the unknown distance between the beacon nodes and target points. The observation equation can be described as.

$$\begin{cases} (x_1 - x)^2 + (y_1 - y)^2 = d_1^2 \\ (x_2 - x)^2 + (y_2 - y)^2 = d_2^2 \\ \quad\vdots \\ (x_M - x)^2 + (y_M - y)^2 = d_M^2 \end{cases} \tag{12}$$

When $M = 3$, the initial coordinate of the target point based on the trilateration positioning is as follows

$$\hat{X}_0 = \begin{bmatrix} 2(x_1 - x_2) & 2(y_1 - y_2) \\ 2(x_1 - x_3) & 2(y_1 - y_3) \end{bmatrix}^{-1} \cdot \begin{bmatrix} x_1^2 - x_2^2 + y_1^2 - y_2^2 + d_2^2 - d_1^2 \\ x_1^2 - x_3^2 + y_1^2 - y_3^2 + d_3^2 - d_1^2 \end{bmatrix}$$

When $M > 3$, the estimated coordinate $\hat{X} = (\hat{x}, \hat{y})^T$ is generally used to replace the real coordinates $X = (x, y)^T$ of the target point in the LS algorithm. $L = (d_1^2, d_2^2, \ldots, d_m^2)^T$ is the distance vector, $V = (v_x, v_y)^T$ is the correction vector of the target point coordinate. $f(\hat{X}) = (f_1(\hat{X}), f_2(\hat{X}), \ldots, f_M(\hat{X}))^T$ is the vector representation of the non-linear function of the observation equation. $f(\hat{X})$ is expanded to the primary term by Taylor expansion [14] at $\hat{X}_0$. The non-linear error equation is obtained.

$$V = Bd\hat{X} - \left(L - f\left(\hat{X}\right)\right) \tag{13}$$

where B is the Jacobian matrix of $f(\hat{X})$ at $\hat{X}$.

$$\begin{cases} B = \dfrac{\partial f\left(\hat{X}\right)}{\partial \hat{X}} \\ d\hat{X} = \hat{X} - X \end{cases}$$

The coordinate of the target point is then obtained according to the nonlinear LS algorithm.

$$\begin{cases} d\hat{X} = \left(B^T B\right)^{-1} B^T \left(L - f\left(\hat{X}\right)\right)\Big|_{\hat{X} = \hat{X}_0} \\ \hat{X}_1 = X_0 + d\hat{X} \end{cases} \tag{14}$$

3.2 Gauss-Newton Iterative Algorithm

It should be note that the LS algorithm cannot obtain a high-precision positioning result when a large error in RSSI ranging. Therefore, $\hat{X}_1$ as the approximate coordinate of the target point, the position determination is obtained by Gauss-Newton iterative algorithm. The $(L - f(\hat{X}_p))$ is calculated by using the coordinates $\hat{X}_p$ obtained at the p-th time. At the same time, the physical distance between the target point and the beacon node is weighted. The weight coefficient is determined by the pseudo-range difference. This method requires the distance between the target point and the beacon node minus the distance between the RP and the same beacon node. Compared with the traditional inverse physical distance weighted method, the advantage of proposed is that only one model parameter is used. So, it is easy to get the difference equation according to formula (2).

$$PD_i^j = \left| 10^{\wedge} \left(\frac{P_r\left(d_{rp}^j\right)_{dBm} - P_r(d_0)_{dBm}}{10\gamma} \right) - 10^{\wedge} \left(\frac{P_r\left(d_i^j\right)_{dBm} - P_r(d_0)_{dBm}}{10\gamma} \right) \right|$$

$$= 10^{\wedge} \left(\frac{-P_r(d_0)_{dBm}}{10\gamma} \right) \cdot \left| 10^{\wedge} \left(\frac{P_r\left(d_{rp}^j\right)_{dBm}}{10\gamma} \right) - 10^{\wedge} \left(\frac{P_r\left(d_i^j\right)_{dBm}}{10\gamma} \right) \right| \tag{15}$$

where $P_r\left(d_i^j\right)_{dBm}$ indicates the signal strength between the i-th target point and the j-th beacon node. $P_r\left(d_{rp}^j\right)_{dBm}$ is the RSSI between the RP and the j-th beacon node. Therefore, the weight coefficient of the i-th target point can be defined as ω_i^j, and the weighting matrix is expressed as $W = diag\left(\omega_i^1, \omega_i^2, \ldots, \omega_i^j, \ldots, \omega_i^M\right)$. The Gauss-Newton iteration formula is shown in (17).

$$\omega_i^j = \frac{PD_i^j}{\sum\limits_{j=1}^{M} PD_i^j}$$

$$= \frac{10^\wedge\left(\frac{-P_r(d_0)_{dBm}}{10\gamma}\right) \cdot \left|10^\wedge\left(\frac{P_r\left(d_{rp}^j\right)_{dBm}}{10\gamma}\right) - 10^\wedge\left(\frac{P_r\left(d_i^j\right)_{dBm}}{10\gamma}\right)\right|}{\sum\limits_{j=1}^{M} 10^\wedge\left(\frac{-P_r(d_0)_{dBm}}{10\gamma}\right) \cdot \left|10^\wedge\left(\frac{P_r\left(d_{rp}^j\right)_{dBm}}{10\gamma}\right) - 10^\wedge\left(\frac{P_r\left(d_i^j\right)_{dBm}}{10\gamma}\right)\right|} \quad (16)$$

$$= \frac{\left|10^\wedge\left(\frac{P_r\left(d_{rp}^j\right)_{dBm}}{10\gamma}\right) - 10^\wedge\left(\frac{P_r\left(d_i^j\right)_{dBm}}{10\gamma}\right)\right|}{\sum\limits_{j=1}^{M} \left|10^\wedge\left(\frac{P_r\left(d_{rp}^j\right)_{dBm}}{10\gamma}\right) - 10^\wedge\left(\frac{P_r\left(d_i^j\right)_{dBm}}{10\gamma}\right)\right|}$$

$$\begin{cases} d\hat{X}_p = \left(B^T W B\right)^{-1} B^T W \left(L - f\left(\hat{X}\right)\right)\Big|_{\hat{X}=\hat{X}_p}, (p \geq 1) \\ \hat{X}_{p+1} = \hat{X}_p + d\hat{X}_p \end{cases} \quad (17)$$

Terminate the iteration when $(V^T W V)^{p+1} = (V^T W V)^p$.

3.3 Accuracy Measure

The accuracy measure in Gauss-Newton iterative positioning is determined by physical distance between the real coordinates and the estimated coordinates of the target point, that is

$$error = \left\|\hat{X}_p - X\right\| \quad (18)$$

4 Experimental Results

In the experiment, we use the real measurement data of "UPINLBS2019 Indoor Localization Competition" to evaluate the performance of the proposed algorithm. The field experiment area and data acquisition process are as follows: multiple beacon nodes and target points are deployed in the corridor of 65 m × 3 m. The relative coordinates between them are obtained by measurement technology, and a target point with known coordinates is taken as the reference point (RP). Each target point can receive the wireless signals from all beacon nodes and record them in the (.txt) file. Four of them (i.e. $M = 4$) with the maximum RSSI preprocessing value were selected for good precision positioning. A specific beacon node is used to selects to establish the CTM, the sampling interval of 0.3 m was then used to gather RSSI for about 50 groups.

4.1 BLE Signal Processing

The BLE signal processing algorithm proposed is used to process the RSSI when the distance between the target point and beacon node is 3.0 m. The results are shown in Fig. 1. After the algorithm proposed, the error between the wireless signal and the real signal is smaller.

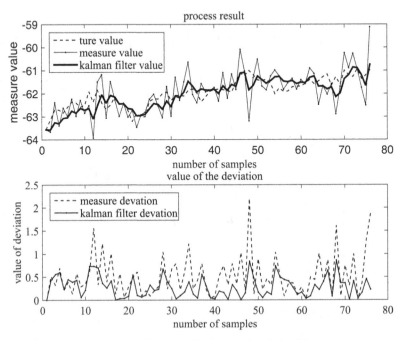

Fig. 1. RSSI processed by the proposed algorithm

The proposed algorithm is used to process the 50 sets RSSI observations. And then the linear regression algorithm is used to fit the functional relationship between RSSI and distance, which can determine the parameters $Pr(d_0)_{dBm}$ and γ. The function fitting accuracy is improved by gradually eliminating abnormal signals with errors exceeding 1 dBm. The final fitting result is shown in Fig. 2, and parameter $Pr(d_0)_{dBm} = -56.1605$, $\gamma = 2.2848$.

The different signal strength corresponding distance was obtained by taking the model parameters $Pr(d_0)_{dBm}$ and γ into formula (2). For evaluating the performance of the parameter estimation algorithm, experiments were conducted comparing with the actual distance. The comparison error is shown in Fig. 3.

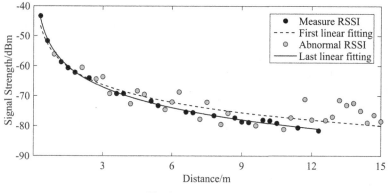

Fig. 2. Curve fitting

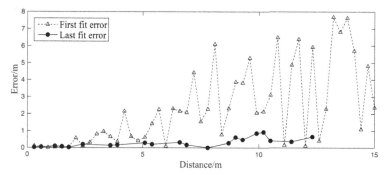

Fig. 3. Distance error

The errors of the first parameter estimation are between 0–8 m. The second parameter estimation is performed by deleting the abnormal signal whose error exceeding 5 dBm, and the third parameter estimation is performed by deleting the abnormal signal whose error exceeding 4 dBm. Until the abnormal signal with an error exceeding 1 dBm was deleted, the sixth parameter estimation is performed. The error between the calculated distance and the actual distance was maintained within 1 m, and the average error was about 0.3 m. The experimental results demonstrate that the model parameter estimation algorithm can effectively improve the parameter accuracy. Meanwhile, it has obvious advantages in improving the RSSI ranging accuracy.

4.2 Comparison of Positioning Accuracy

In order to verify the effectiveness of the proposed algorithm, five different algorithms are conducted in the experiments. The LS algorithm based on MF, GF, KF, Wavelet Kalman filtering (WKF) and proposed. Among them, the first three filtering algorithms used the highest frequency to process the RSSI observations. Empirical parameters are usually used in the CTM model to convert the physical distance. Firstly, the WKF-based LS positioning algorithm uses wavelet threshold denoising and Kalman filtering to remove noise and smooth the RSSI, and then the parameters in the CTM are determined by parameter estimation algorithm. Finally, the coordinates of the target point are calculated according to the LS positioning principle.

The statistical 8 target points positioning results for five different positioning algorithms and literature [15] are displayed in Table 1. The positioning errors of the proposed are generally concentrated within 4 m, and the average error is about 2 m. The positioning accuracy is significantly better than 3.7 m of MF algorithm, 2.8 m of WKF algorithm and 3 m of GF algorithm and KF algorithm. Meanwhile, compared with literature [15], the accuracy of proposed algorithm is improved by 25.63%.

Table 1. Comparison of positioning accuracy of 5 algorithms (m)

Algorithm	Point number								
	No. 1	No. 2	No. 3	No. 4	No. 5	No. 6	No. 7	No. 8	Average error
MF	1.35	9.37	0.64	2.88	5.09	6.02	2.00	2.02	3.70
GF	1.20	6.03	0.81	3.00	4.04	2.03	2.35	3.60	3.01
KF	1.14	6.45	0.83	3.92	4.03	2.04	2.35	3.61	3.05
WKF	1.18	4.84	0.76	3.97	4.01	2.10	2.35	3.54	2.84
Literature [15]	1.30	3.89	0.83	4.01	3.91	2.15	2.52	3.60	2.77
Proposed	1.18	1.68	0.65	1.20	4.00	1.96	2.27	3.54	2.06

The Cumulative Distribution Function (CDF) of five different algorithms is analyzed for in this experiment, the results are shown in Fig. 4.

The results are displayed that the overall positioning error based on BLE-RSSI is stable within 10 m. The proposed can obtain the higher positioning accuracy significantly better than MF algorithm, GF algorithm, KF algorithm, WKF algorithm and literature [15], which explains the Gauss Newton iterative positioning has obvious advantages in improving the positioning accuracy. Although the maximum positioning error of the proposed and literature [15] are maintained at about 4 m, the CDF of proposed is 75% when the positioning error is 3 m. It is significantly higher than 62.5% of the literature [15] and MF algorithm, 50% of the other three algorithms. When the positioning error threshold is 4 m, the CDF of proposed and literature [15] are almost 100.0%, which is higher than 62.5% of the MF algorithm and 87.5% of the others algorithms. In addition, the maximum positioning error of the WKF algorithm is less than 5 m, which indicates that the use of wavelet threshold denoising to process RSSI has a certain effect.

In the case that the process of other algorithms does not change, the experiment sets up a group of weighting method using the inverse of physical distance to compare with the proposed weighting method. The positioning error is displayed in Table 2. The average positioning error weighted by pseudo-range difference is about 2 m, which is less than 2.4 m of the inverse weighting method of physical distance. It is displayed that the weighting coefficient error caused by the CTM parameter error will affect the indoor positioning accuracy. The proposed weighting method can be effectively reduce the weight coefficient error caused by the model parameter error.

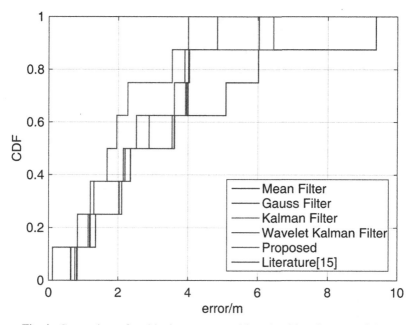

Fig. 4. Comparison of positioning accuracy of five algorithms in terms of CDF

Table 2. Comparison of positioning accuracy of different weighting methods (m)

Algorithm	Point number								
	No. 1	No. 2	No. 3	No. 4	No. 5	No. 6	No. 7	No. 8	Average error
Proposed	1.18	1.68	0.65	1.20	4.00	1.96	2.27	3.54	2.06
inverse weighting	1.61	3.68	0.88	1.46	4.02	1.78	2.27	3.58	2.41

4.3 Position Performance Indexes Comparison

In addition to positioning accuracy, the operating efficiency and robustness of algorithm are all indexes to evaluate the positioning performance. Table 3 displays the operating efficiency and robustness comparison of the proposed, literature [15] and other four different algorithms. The Gauss-Newton algorithm cannot converge caused by the insufficient RSSI ranging accuracy based on mean filtering or Gauss filtering algorithm. Based on Kalman filter algorithm after wavelet threshold denoising, the operation efficiency is increased by 3.73% in the process of Gauss-Newton iterative positioning. The presented average operating time of algorithm is 9.79 s. Compared with the Gauss-Newton iterative localization algorithm based on KF or WKF, the operation efficiency is improved 13.13% and 9.77% respectively. There is no Gauss-Newton process in the literature algorithm, so its operation efficiency is lower than the other five algorithms. But the complexity of literature is higher than other algorithms because of the literature uses a hybrid algorithm of centroid and LS algorithm. Considering that sometimes the LS positioning algorithm has no solution, which is a troublesome process. Meanwhile, the literature has no weighting in the process of LS, it will have a great impact on the positioning results. Compared with the physical distance inverse weighting method, the new weighting algorithm only requires one parameter. Considering that the RSSI processing algorithm proposed can get more stable signal than the MF, GF, KF and WKF algorithm, the robustness of the proposed is stronger than other algorithms.

Table 3. Position performance indexes comparison

Algorithm	MF	GF	KF	WKF	Literature	Proposed
Operating efficiency/s	–	–	11.27	10.85	6.82	9.79
Robustness	Weak	Weak	Weak	Weak	Weak	Strong

5   Conclusion

We proposed a new BLE-RSSI indoor positioning algorithm based on wavelet threshold denoising and pseudo-range differential weighting in this paper. The experiment results demonstrate that processing the RSSI observations by using wavelet threshold denoising and deleting abnormal signals can reduce the effect of signal noise and large disturbance anomalous values on the results of Kalman filtering. Combined with the CTM parameter estimation algorithm to determine model parameters, it can effectively improve RSSI ranging accuracy. The proposed weighting method can use a model parameter γ to obtain a higher precision weight coefficients than traditional physical distance inverse weighting methods. Using Gauss-Newton iterative algorithm to continuously modify the target point deviation can obtain the coordinates of indoor target points. At the end of the experiment, compared with the five positioning algorithms, the average error of proposed is significantly smaller than the other algorithms, indicating that the algorithm of proposed has obvious advantages.

References

1. Wu, X., Shen, R., Fu, L., et al.: iBILL: using iBeacon and inertial sensors for accurate indoor localization in large open areas. IEEE Access **5**, 14589–14599 (2017)
2. Cantón Paterna, V., Calveras Auge, A., Paradells Aspas, J., et al.: A bluetooth low energy indoor positioning system with channel diversity, weighted trilateration and Kalman filtering. Sensors **17**(12), 2927 (2017)
3. Yiu, S., Dashti, M., Claussen, H., et al.: Wireless RSSI fingerprinting localization. Signal Process. **131**, 235–244 (2016)
4. Xu, Y., Liu, H., Ma, Z., et al.: Multipath-tolerant ranging algorithm in underground tunnel for wireless sensor networks. Chin. J. Sci. Instrum. **038**(010), 2461–2468 (2017)
5. Ni, X., Gao, Y., Li, L.: Hybrid filtering algorithm based on RSSI. Comput. Sci. **46**(8), 133–137 (2019)

6. Li, Y., Zhuang, Y., Lan, H., et al.: A hybrid WiFi/magnetic matching/PDR approach for indoor navigation with smartphone sensors. IEEE Commun. Lett. **20**(1), 169–172 (2016)
7. Xue, W., Hua, X., Li, Q., et al.: A new weighted algorithm based on the uneven spatial resolution of RSSI for indoor localization. IEEE Access **6**, 26588–26595 (2018)
8. Zhou, C., Yuan, J., Liu, H., et al.: Bluetooth indoor positioning based on RSSI and Kalman Filter. Wireless Pers. Commun. **96**(3), 1–16 (2017)
9. Sari, R., Zayyani, H.: RSS localization using unknown statistical path loss exponent model. IEEE Commun. Lett. **22**(9), 1830–1833 (2018)
10. Shen, L.L., Hui, W.W.S.: Improved pedestrian dead-reckoning-based indoor positioning by RSSI-based heading correction. IEEE Sens. J. **16**(21), 7762–7773 (2016)
11. Zhang, B., Wang, H., Zheng, L., et al.: Joint synchronization and localization for underwater sensor networks considering stratification effect. IEEE Access **5**, 26932–26943 (2017)
12. Zhu, M., Lu, X., Lu, Z., Li, Y., Tao, X.: RSSI indoor ranging algorithm combining wavelet transform and neural network. Bull. Surv. Mapp. **1**, 50–54 (2020)
13. Fang, X., Nan, L., Jiang, Z., et al.: Multi-channel fingerprint localisation algorithm for wireless sensor network in multipath environment. IET Commun. **11**(15), 2253–2260 (2017)
14. Fengjun, S., Yi, J., Anping, X., et al.: A node localization algorithm based on multi-granularity regional division and the lagrange multiplier method in wireless sensor networks. Sensors **16**(11), 1934 (2016)
15. Ni, Y.F., Shi, X.H.: Indoor staff Kalman filter location algorithm based on RSSI. J. Xi'an Univ. Sci. Technol. **40**(1), 167–172 (2020)

Wetland Classification Using Sparse Spectral Unmixing Algorithm and Landsat 8 OLI Imagery

Jie Ding[1], Xiaodong Na[1(✉)], and Xingmei Li[2]

[1] Heilongjiang Province Key Laboratory of Geographical Environment Monitoring and Spatial Information Service in Cold Regions, Harbin Normal University, Harbin 150025, China
[2] Faculty of Mechanical and Electronics Engineering, China University of Geosciences, Wuhan 430074, China

Abstract. Linear spectral unmixing algorithm has been widely applied in estimating land cover abundances within a wetlands distributed area. However, researchers have to examine the appropriateness of endmemebers in a large spectra library, and may select the ones that could not exist in the particular pixel by mistake. Such nonexistent endmember signatures may lead to the decreasing of classification accuracies. In the current study, we proposed a sparse unmixing algorithm to discriminate the distribution of wetlands in a cold region based on Landsat 8 OLI multispectral images, and compared the classification accuracies with that from the traditional linear unmixing algorithm. Considered the spatial heterogeneity of endmembers spectra, the sparse spectral unmixing algorithm adaptively selects the combination of endmembers for each pixel, and compensates the decencies of the traditional linear unmixing method during the process of endmembers selections. Experimental results suggest that the sparse spectral unmixing algorithm is with apparently better performance when compared to those of linear spectral unmixing algorithm, signified with lower root-mean-square error and higher correlation. Therefore, the sparse unmixing algorithm showed a significantly potential in increasing the wetland classification accuracies and precisions.

Keywords: Landsat-8 OLI · Sparse constrained · Linear spectral unmixing · Wetland classification

1 Introduction

Wetland is an important ecosystem on earth, which is of great significance in terms of species storage, climate regulation, ecological balance maintenance and biodiversity conservation [1]. However, under the heavy effects of natural and human disturbances, the global wetlands have demonstrated a serious tendency of degradation and disappearance [2]. Obtaining the accurate wetlands spatial distribution information through investigating and monitoring technologies are the basis of wetlands protection and management. During the recent years, the wetland monitoring methods have developed from field comprehensive survey to dynamic monitoring using GIS and RS techniques. The

G. Pan et al. (Eds.): SpatialDI 2021, LNCS 12753, pp. 186–194, 2021.
https://doi.org/10.1007/978-3-030-85462-1_17

monitoring objectives have also expanded from the wetland recognition, wetland area calculating to researches on wetland vegetation communities, spatial distribution pattern and species diversity [3]. However, the wetland monitoring and mapping were extremely difficult due to its inaccessible, high spatial heterogeneity, and sophisticate hydrological characteristics [4, 5].

During the past decades, multi-source satellite images from different sensor platforms have become the main data sources for wetland information extraction and dynamic monitoring due to its advantages of large scale and repeat observations. Especially, the multispectral sensors such as Landsat (landsat-5 TM, landsat-7 ETM+, and Landsat 8 OLI) and SPOT (SPOT-5, SPOT-6 and SPOT-7) have been successfully used on wetland mapping and monitoring [6, 7]. In addition, many effective machine learning algorithms have been proposed to recognize the wetlands, including pixel-oriented classification algorithms (maximum likelihood classification, support vector machine, decision trees, and random forest, etc.) [3, 8], object-oriented classification algorithms [9, 10], and spectral unmixing algorithms [11]. In terms of medium and low spatial resolution remote sensing images, the spectral unmixing algorithms are preferred because each pixel in these images may contain multiple components influenced by the sophisticate structure of the wetland vegetation communities. However, the traditional linear spectral unmixing algorithm uses the same endmembers for all pixels in one scene, and neglects the spatial variability of the endmember spectrum, which results in the misclassification for the sophisticate components of wetlands. One potential solution to this problem is to incorporate a priori knowledge of endmember distribution. For generating such a priori knowledge, this paper resorts to sparse unmixing algorithm. Sparse unmixing algorithm assumes that the number of "active" endmembers (actually contained in the mixed pixel) is very small, and the abundances of most endmembers in a particular mixed pixel are zero or approach zero [12, 13]. On the basis of the Abundance Nonnegative Constraints (ANC), Abundance Sum to one Constraints (ASC), as well as the Sparsity Constraints (SC) of the fractional abundance vectors inherited in the sparse unmixing algorithm, the "active" endmembers existing in each mixed pixel could be adaptively selected. Then these selected endmembers are used as a priori conditions to estimate the fractional abundance of each pixel in a sophisticate scene.

In the current study, sparse unmixing algorithm was used to extract the spatial distribution information of the wetlands in Nanweng River Naional Natural Reserve, Da Hinggan Mountains, Northeast of China using Landsat-8 OLI images. The classification accuracies of the sparse unmixing algorithm were compared with the traditional linear unmixing algorithm to testify the feasibility of the proposed method.

2   Methodology

2.1   Sparse Spectral Unmixing Algorithm

Sparse unmixing assumes that only a small number of endmembers existed within a mixed pixel, although with the availability of a large spectral signature library [14]. It is a semi-supervised spectral unmixing method, and the corresponding unmixing model can be written as follows:

$$\min_{x} \|Ax - y\|_F^2 + \lambda \|x\|_2, x \geq 0, \sum_{i=1}^{m} |x_i| = 1 \qquad (1)$$

Where y is the spectral feature values of the Landsat-8 OLI images; A is a endmembers spectrum library; $x \geq 0$ represents the Abundance Nonnegative Constraints (ANC) of the fractional abundance vectors; $\sum_{i=1}^{m} |x_i| = 1$ represents the Abundance Sum to one Constraints (ASC) of the fractional abundance vectors; $\|X\|_F \equiv \sqrt{trace(XX^T)}$ is the Frobenius norm; The regularization parameter $\lambda > 0$ is a regularization parameter which was used to weight the fitness and the sparsity. $\|x\|_2$ is the L_2 norm, which represents the number of nonzero in the fractional abundance vectors, and enforces joint sparsity to all pixels. The sparsity feature of the fractional abundance values guarantee that most nonexistent endmembers in the current pixel are assigned zeros or approached zeros. Therefore, we can select the "active" endmembers and their corresponding spectral signatures through the Sparsity Constraints (SC) of the fractional abundance vectors. After that, the fractional abundance values were calculated for each pixel according to the selected "active" endmembers and their signatures.

2.2 Linear Spectral Unmixing Algorithm

The traditional linear spectral unmixing model can be expressed as follows:

$$y_i = \sum_{m=1}^{M} C_{i,m} x_m + n_i, \sum_{m=1}^{M} x_m = 1 \qquad (2)$$

Where y_i is the spectral signature of a mixed pixel at band i of Landsat-8 OLI images, $C_{i,m}$ is the spectral signature of the mth endmember at the band i, x_m is the corresponding abundance value of the mth endmember, noise n_i represents the error term of band i, and M represents the size of the endmembers library. The traditional linear spectral unmixing algorithm was conducted to compare the classification accuracies with the results of sparse spectral unmixing algorithm.

3   Simulation

3.1   Study Area

Nanweng River wetlands are located at the southern of the greater Khingan Mountains, Heilongjiang province, China, with a latitude between 51°05′ N and 51°40′ N, and longitude between 125°07′ E and 125°50′ E (see Fig. 1). The Nanweng River National Nature Reserve was listed as a "Wetland of International Importance" in 2011 by the Ramsar convention and became an internationally important wetland. The Nanweng River National Nature Reserve is in a cold temperate continental monsoon climate zone with a total area of 2295 km². The primary land cover types are marsh, meadow, forest, lake, and dry land. The marsh vegetation communities are composed of herbaceous wetlands, forest wetlands, shrub wetlands, island forest wetlands, and lake wetlands. The unique cold temperate forest wetland landscape and rich biodiversity of Nanweng River wetlands provide a valuable study area for wetland monitoring and protection.

Fig. 1. Location of the study area.

3.2 Data Preparation

One cloud-free Landsat-8 OLI image was obtained from the Northeast Institute of Geography and Agricultural Ecology, Chinese Academy of Science, Changchun, China. The image was acquired on 14 June 2015. Seven multispectral bands with 30 m spatial resolution (band 1 to 7) and one panchromatic band with 15 m spatial resolution (band 8) were used in this study. The image was georeferenced by 66 ground control points derived from 1:50000 topographic maps. The geo-correction procedure achieved a root-mean-square error (RMSE) less than 0.5 pixels. The geoeferenced images were atmospheric corrected based on FLAASH tool in the ENVI 5.1 software. The Landsat-8 OLI panchromatic band were merged with OLI 1-7 bands through Gram-Schmidt algorithm to improve both the spatial resolution and spectral resolution. At last, the merged images were clipped based on the vector boundary of Nanwenghe River National Nature Reserve. The endmembers spectral library was constructed based on the spectral features of six land cover types, including marsh, meadow, forest, residential land, and cultivated land. The spectral features of endmembers were selected directly from the Landsat-8 OLI images using the minimum noise fraction transformation (MNF) and pure pixel index (PPI) technologies provided by ENVI 5.1 software.

3.3 Abundance Estimation Using a Sparse Unmixing Algorithm

With the Landsat-8 OLI generated from Sect. 3.2, we estimate the abundance of six endmembers components in each pixel using a sparse unmixing algorithm which developed by the MATLAB 2018b software. During the process of determining spectral signatures, the developed sparse unmixing with l_2 norm was applied to choose the "active" endmembers for each pixel. The abundance threshold T was assigned to 0.01 in order to select the "active" endmembers. If the abundance of an endmember is less than the threshold T, it would be eliminated for the further spectral unmixing process. In the process of sparse unmixing, the regularization parameter λ is empirically set as 0.01, and the abundances of six land cover types were estimated accordingly. The abundance maps of marsh, meadow, forest, and cultivated land were shown in Fig. 2 (see Fig. 2).

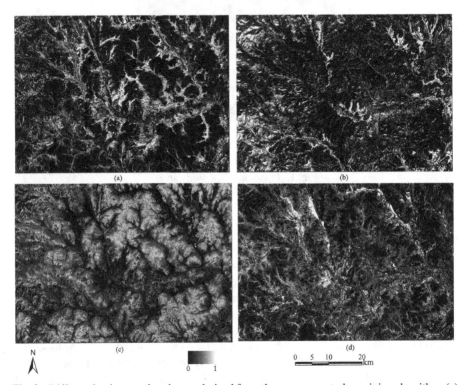

Fig. 2. Different land cover abundances derived from the sparse spectral unmixing algorithm. (a) Marsh, (b) Meadow, (c) Forest, and (d), Cultivated land.

3.4 Accuracy Assessment and Comparative Analysis

High Resolution Satellite Images from Google Earth were used to assess the spectral unmixing results. Firstly, 50 testing plots were randomly selected from the high resolution satellite images with a 30 × 30 rectangular window (30 m × 30 m). The land land cover types of the 50 testing plots were recognized by the visual interpretation method. The abundances of different land cover types in each testing plot were calculated and compared with the abundances of spectral unmixing. The correlation coefficient (R) and root-mean-square error (RMSE) were utilized to assess the accuracies of abundances estimation.

$$R = \frac{\sum_{i=1}^{N} \left(X_i - \overline{X}\right)\left(Y_i - \overline{Y}\right)}{\sqrt{\sum_{i=1}^{N} \left(X_i - \overline{X}\right)^2}\sqrt{\sum_{i=1}^{N} \left(Y_i - \overline{Y}\right)^2}} \tag{3}$$

$$RMSE = \left[\sum_{i=1}^{N} (X_i - Y_i)^2 / N\right]^{1/2} \tag{4}$$

Where X_i represents the estimated value of the i sample, X is the estimated mean value, Y_i represents the verification value of the i sample, and Y represents the verification mean. N is the number of samples.

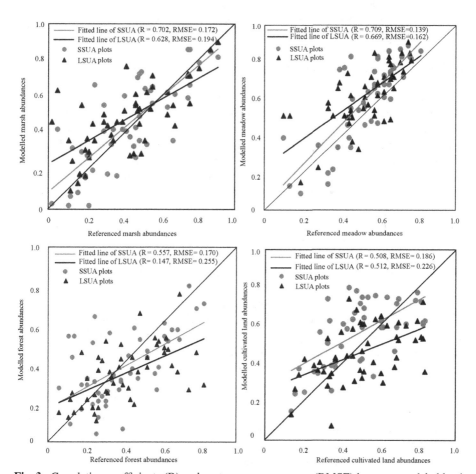

Fig. 3. Correlation coefficients (R) and root mean square error (RMSE) between modeled land cover abundances and referenced land cover abundances retrieved by the sparse spectral unmixing algorithm (SSUA) and the linear spectral unmixing algorithm (LSUA).

The accuracies of both sparse ummixing and linear ummixing models were shown in Fig. 3. Analyses of results suggest that the sparse ummixing method has better performance when compared with the traditional linear spectral unmixing method. For example, for almost all land cover types, the RMSE of the sparse ummixing model is lower than that of linear unmixing model. As was shown in Fig. 3.

The sparse ummixing model improved the accuracies and the fitness of the linear ummixing model. This was mainly due to the different endmember selection methods for the two spectral ummixing models. The traditional linear ummixing model calculates the abundances based on the same endmembers combinations, and all endmembers participate in the spectral unmixing process. While sparse unmixing model eliminates the nonexistent endmembers from the spectral library based on sparsity constraints, and adaptively selects the potential endmembers to calculate the abundances for each pixel. Therefore, the proposed method avoids the impacts of the participation of nonexistent endmembers on the spectral unmixing accuracies and improves the wetland classification accuracies in the sophisticated scene at the same time.

4   Discussion and Conclusions

In the current study, the proposed sparse unmixing algorithm was used to map the wetlands with a commonly used multispectral remote sensing data source, the Landsat-OLI images. Taking the visual classification results from high-resolution images as testing data, we compared the classification accuracies of sparse unmixing algorithm with that from the traditional linear spectral unmixing algorithm. The feasibility of wetland mapping based on sparse unmixing algorithm was quantitatively testified and the impact factors of uncertainty were discussed.

The accuracy comparison between estimated and testified abundances of different land cover obtained from SSUA and LSUA showed that the proposed SSUA model improved the accuracies of the marsh abundances compared with the traditional LUSA (Fig. 3). This was largely due to the different endmembers selection modes for the above spectral unmixing algorithms. The traditional LUSA algorithm used fixed endmmebers to estimate the abundances for each pixel. Although many studies have testified that there is an advantage of spectral unmixing algorithm to estimate vegetation abundances, the spectral spatial heterogeneous of the endmembers prevent the effectiveness of the traditional spectral unmixing algorithm with the fixed endmembers in the whole image, especially to recognize vegetation communities with the similar spectral signature in a sophisticate scene such as the wetland distributed area [15]. Under the influences of high spectral spatial heterogeneity in complex scenes, the participation of the fixed endmembers usually leads to the overestimation of the nonexistent endmember. While the SSUA could adaptively select the optimal endmembers for each pixel based on sparsity constraints. Through compressing the priority probability of the nonexistent endmember, the SSUA improved the estimated accuracies of different land covers abundances in a sophisticate scene. Therefore, the proposed method avoids the impacts of nonexistent endmembers and improves the wetland unmixing accuracies in the sophisticate scene.

Although we have testified the feasibility of the sparse spectral unmixing algorithm, there are also some problems should be concerned during the process of classification. We did not select the endmembers directly from the field gauged spectrum due to the unmatched spectrum between the handheld spectroradiometer and satellite sensors. Appropriate methods should be proposed to perform the spectral unmixing algorithm based on the endmembers selected from field gauged spectrum. Although the wetlands and their surrounding land cover types could be recognized in the current study, the limited spectrum of multi-spectral images could not discriminate the wetlands vegetation communities. The future work would introduce hyperspectral images to classify the wetland dedicatedly, and improve the feasibility of the sparse spectral unmixing algorithm.

The main conclusions are as follows: Compared with the linear spectral unmixing algorithm, the sparse spectral unmixing algorithm improved the wetland mapping accuracies. The correlation coefficient between modelled values and actual values of the fractional marsh and meadow from sparse spectral unmixing algorithm were all above 0.8, and the RMSE were lower than 0.1, which testified that the proposed spectral unmixing algorithm was suitable to map the typical freshwater marsh wetlands in the cold regions. In addition, the sparse spectral unmixing algorithm could select the endmembers adaptively. The selected endmembers were used to spectral unmixing for each pixel which relieves the spatial heterogeneity of endmembers and improve the classification accuracies of wetlands and its surrounding land cover types.

Acknowledgement. This research was funded by Natural Science Foundation of Heilongjiang Province, China, grant number YQ2020D005 and Heilongjiang province innovation and entrepreneurship for college students, grant number 202010231080.

References

1. Daneshva, F., Nejadhashemi, Woznicki, A.P., Sean, A.W., Matthew, R.H.: Applications of computational fluid dynamics in fish and habitat studies. Ecohydrol. Hydrobiol. **17**(1), 53–62 (2017)
2. Rt Park, H., Kim, Y., Kimball, J.S.: Widespread permafrost vulnerability and soil active layer increases over the high northern latitudes inferred from satellite remote sensing and process model assessments. Remote Sens. Environ. **175**(15), 349–358 (2016)
3. Na, X.D., Zang, S.Y., Zhang, Y.H., Li, W.L.: Assessing breeding habitat suitability for the endangered red-crowned crane (Grus japonensis) based on multi-source remote sensing data. Wetlands **35**, 955–967 (2015)
4. Zhang, M., Gong, Z., Zhao, W., Pu, R., Liu, K.: Estimating wetland vegetation abundance from Landsat-8 operational land imager imagery: a comparison between linear spectral mixture analysis and multinomial logit modeling methods. J. Appl. Remote Sens. **10**(1), 015005 (2016)
5. Na, X.D., Zang, S.Y., WU, C.S., Yang, T., Li, W.L.: Hydrological regime monitoring and mapping of the Zhalong Wetland through integrating time series Radarsat-2 and Landsat imagery. Remote Sens. **10**(5), 702 (2018)
6. Han, X., Chen, X., Feng, L.: Four decades of winter wetland changes in Poyang Lake based on Landsat observations between 1973 and 2013. Remote Sens. Environ. **156**, 426–437 (2015)
7. Taddeo, S., Dronova, I., Depsky, N.: Spectral vegetation indices of wetland greenness: responses to vegetation structure, composition, and spatial distribution. Remote Sens. Environ. **234**(1), 111467 (2019)

8. Ma, T.X., Song, X.F., Svenning, C., Li, R.K.: Linear spectral unmixing using endmember coexistence rules and spatial correlation. Int. J. Remote Sens. **39**(11), 3512–3536 (2018)

9. Na, X.D., Zang, S.Y., Wu, C.S., Li, W.L.: Mapping forested wetlands in the Great Zhan River Basin through integrating optical, radar, and topographical data classification techniques. Environ. Monit. Assess. **187**(11), 1–17 (2015)

10. Onojeghuo, A.O., Onojeghuo, A.R.: Object-based habitat mapping using very high spatial resolution multispectral and hyperspectral imagery with LiDAR data. Int. J. Appl. Earth Obs. Geoinf. **59**, 79–91 (2017)

11. Bian, J.H., et al.: Monitoring fractional green vegetation cover dynamics over a seasonally inundated alpine wetland using dense time series HJ-1A/B constellation images and an adaptive endmember selection LSMM model. Remote Sens. Environ. **197**, 98–114 (2017)

12. Iordache, M.D., Bioucas-Dias, J.M., Plaza, A.: Sparse unmixing of hyperspectral data. IEEE Trans. Geosci. Remote Sens. **49**(6), 2014–2039 (2011)

13. Yue, J.B., Tian, Q.J., Tang, S.F., Xu, K.J., Zhou, C.Q.: A dynamic soil endmember spectrum selection approach for soil and crop residue linear spectral unmixing analysis. Int. J. Appl. Earth Obs. Geoinf. **78**(6), 306–317 (2019)

14. Bioucas-Dias, J.M., et al.: Hyperspectral unmixing overview: geometrical, statistical, and sparse regression-based approaches. IEEE J. Select. Topics Appl. Earth Obs. Remote Sens. **5**(2), 354–379 (2012)

15. Chen, B., Chen, L.F., Lu, M., Xu, B.: Wetland mapping by fusing fine spatial and hyperspectral resolution images. Ecol. Model. **353**(10), 95–106 (2017)

City Analysis

A Fine-Grained Mixed Land Use Decomposition Method Based on Multi-source Geospatial Data

Chunyang Huang, Xun Liang, Zhijiang Yang$^{(\boxtimes)}$, and Qingfeng Guan

School of Geography and Information Engineering, China University of Geosciences,
Wuhan 430074, China

Abstract. The urban mixed land use strategy can improve the efficiency and value of land use, which is of great significance to the sustainable development of cities. Previous studies related to urban land use mainly focused on neighborhood, grid, or parcel scales. These spatial scales are not fine enough to cover the spatial structure information of the internal multi-functional mixture, which is difficult to meet the needs of fine urban planning and construction. To this end, this paper proposes an urban mixed land use decomposition model at the pixel scale with higher spatial resolution, combined with a spectral unmixing strategy. Firstly, this paper constructs a multi-source feature set to describe the natural and socio-economic characteristics of urban functions. Then, the Fully Constrained Least Squares (FCLS) model was used to extract the information of urban functional abundance. The results of the mixed decomposition were compared qualitatively and quantitatively with previous studies, and the overall accuracy was verified to be 0.833 and 0.763 for Kappa. Finally, Shannon Diversity Indicator (SHDI) was constructed to characterize the multifunctional mixed-use level of construction land units.

Keywords: Urban mixed land use · Mixed decomposition · Spectral unmixing

1 Introduction

Mixed land use studies are particularly important in reducing land planning costs, improving the efficiency, frequency, and value of land use, and supporting fine-grained land resource allocation and inventory planning [1]. However, in previous studies on urban land use, most of them reckon that the land use in urban parcels has homogeneity and single land classification labels, and researchers often ignore the mixed nature of land types [2]. To explore the spatial structure and distribution of mixed land use, it is essential to acquire qualitative and quantitative information about mixed land use. In terms of spatial scale of previous urban mixed land use studies, most of them are based on large spatial scales, such as urban road network segmentation unit [3], planning parcels [4], large-scale grids [5], multi-functional buildings [6] to reveal the spatial structure of land use. These studies based on large-scale units ignore the detailed spatial distribution of internal features, and can't meet the needs of urban planning.

© Springer Nature Switzerland AG 2021
G. Pan et al. (Eds.): SpatialDI 2021, LNCS 12753, pp. 197–204, 2021.
https://doi.org/10.1007/978-3-030-85462-1_18

The quantization of mixed urban land is a challenging task, which is often closely related to the physical information, social attribute economy, and human activity information [7]. Difficulties in obtaining accurate ground truth data are the main factors that make this task challenging [8]. To solve this problem, scholars try to use multi-source geospatial data, which contains rich information on social semantics and human activities, to overcome the limitation of traditional remote sensing data in providing information about socio-economic attributes [9].

While previous studies have improved urban land use classification by integrating multi-source geographic data, there are still some shortcomings. On the one hand, most of them do not consider the complex multifunctional mixed nature of urban land use, they instead concern about defining specific land-use types [10]. On the other hand, many papers focus on urban parcels, and the spatial scale is not fine enough to meet the needs of fine management and planning.

Traditionally, spectral unmixing based on remote sensing images is a method to decompose and extract the mixed components of ground objects [11]. It is rather more difficult to extract information on the socio-economic functions of land use (e.g., residential, industrial, commercial, public services, etc.) [12]. Although the proportion of mixed land use within the city can be extracted by combining big data and machine learning algorithms, most previous studies are parcel-based and have not achieved the decomposition of socioeconomic functional land use structure at a finer resolution. But even though these studies still lacked a finer structural decomposition of socio-economic functional land use like pixel-scale, these works contributed new thoughts to solve this problem.

In general, most relevant studies and existing methods for extracting mixed land use structures can barely fulfill the demands of progressively finer urban spatial governance. It is an urgent need for urban planning to measure the urban mixed functional land-use finely and quantitatively. Therefore, based on multi-source geospatial data and Landsat imagery, this paper attempt to explore the multifunctional abundance components by using the spectral unmixing method at a higher-resolution pixel scale.

2   Research Area and Data

The research area selected in this paper is Shenzhen, Guangdong Province, China, as shown in Fig. 1. Situated south of the Tropic of Cancer, ranging from $113°43'$ to $114°38'$ E to $22°24'$ to $22°52'$ N. From the point of view of research value, Shenzhen is a model of planning urban sustainable development. Exploring the structure of the urban land use pattern of this international metropolis has important reference significance for the planning and development of other cities.

Landsat-8 satellite remote sensing image of Shenzhen was used in this study. The data was obtained from the USGS official website (https://www.usgs.gov/). The POI dataset was obtained from the public website (https://lbs.amap.com/). Each POI has attribute fields with name, address, latitude, longitude, and category. The dataset was collected in 2018 and can reflect the latest situation of urban multi-functional land use. The road network dataset was obtained in 2019, including highway, provincial, county, town, residential roads, etc. The roads network data set is an open-source, convenient and detailed dataset, which has been widely used in urban research.

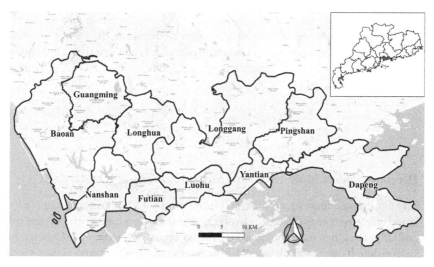

Fig. 1. Overview of the study area

3   Research Methods

3.1   Construction of Multi-source Data Spectral Sets

Specifically, we preprocess the original multi-source data with masking and other operations according to the study scope. Then, every single source data feature is deemed as a simulated urban functional spectral band, and the big data spectral sets are synthesized through the stacking of features.

For remote sensing images, the spatial attributes of ground objects are mainly described by the spectral features and texture features. For a n bands image, spectral features based on the grey value can be described as $spectral = \{spe_1, spe_2, \ldots, spe_n\}$. Texture feature based on Haralick's Gray Level Co-occurrence Matrix (GLCM) set of texture measures [13]. Eight GLCM texture feature factors were calculated texture information, including Mean, Variance, Homogeneity, Contrast, Dissimilarity, Entropy, Second moment, Correlation.

For POI data, Kernel Density Estimation (KDE) can effectively express the distribution density features of POI, thus reflecting the information of urban land use [14]. To be specific, the calculation of KDE method is shown in formula (1), where $\lambda(s)$ is the kernel density calculation function at the spatial position s, r is the distance search radius of reference point i, the distance between s and i is x_{is}, and it less than or equal to r, and n is the total number of points contained in the range x_{is}. K is usually represented as a spatial weight function of distance x_{is} and radius r.

$$\lambda(s) = \sum\nolimits_{i=1}^{n} \frac{1}{\pi r^2} K\left(\frac{x_{is}}{r}\right) \tag{1}$$

At the same time, traffic distance features are closely related to urban land use planning and development patterns, and the distance features between urban ground objects and traffic routes are extracted based on Euclidean distance.

Moreover, after the various features have been extracted, each feature needs to be standardized to eliminate the impact of differences in the scope. As shown in formula (3), the normalized function $Nor(\theta)$ of all the eigenvalue l_θ characteristics are obtained by the Max-Min normalization method, where $max(l)$ and $min(l)$ are the maximum and minimum values respectively.

$$Nor(\theta) = \frac{l_\theta - min(l)}{max(l) - min(l)} \tag{2}$$

3.2   Decomposition of Multi-functional Mixed Land-Use at Fine-Grained

The fully constrained least squares (FCLS) linear mixed decomposition model is more simple and effective than the neural network, matrix decomposition principle, and other methods [15]. Thus, this paper proposed the mixed decomposition method based on the FCLS algorithm. It can be simply described as follows: for multi-source dataset I with n characteristic bands, the features at any discrete spatial coordinate (i, j) in I is represented by a vector $Vec_{X(i,j)} = [x_1(i, j), x_2(i, j), \ldots, x_n(i, j)] \in R$, in which R represents the constituted mixed spectral set. Decomposition of linear mixed model expression of formula (3), in which the E_t denotes the spectral characteristics of the pure endmember t, $\Phi_t(i, j)$ is the abundance function of t at (i, j), m is the total number of endmembers, and $\Delta(i, j)$ is the noise value. For the FCLS model, the abundance results meet the two constraint conditions, one is the Abundance Non-negativity Constraint (ANC), formula expressed as $\Phi_t(i, j) \geq 0$, the second constraint condition is Abundance Sum-to-one Constraint (ASC), formula expressed as $\sum_{t=1}^{m} \Phi_t(i, j) = 1$. The optimization objective of the FCLS model goal is to minimize the gap between the vector $Vec_{X(i,j)}$ and $\Phi_t(i, j) \cdot E_t$, as shown in formula (4).

$$Vec_{X(i,j)} = \sum_{t=1}^{m} \Phi_t(i, j) \cdot E_t + \Delta(i, j) \tag{3}$$

$$min[Vec_{X(i,j)}, \Phi_t(i, j) \cdot E_t] \tag{4}$$

4   Case Experiment

4.1   Result Validation

We randomly sampled 600 grid parcels from the reference data as a verification, with parcels for Commercial, Public, Residential, and Work functional. From the validation statistics table (Table 1), we know that the overall accuracy (OA) and Kappa coefficient of the validation are 0.833 and 0.763 separately. The validation indicators producer accuracy (PA), and user accuracy (UA) for the Public, Residential and Industrial functions mostly over 0.8, and these were highly accurate.

Table 1. Results of precision statistical.

Indicator	Commercial	Public	Residential	Work
PA	0.833	0.812	0.815	0.838
UA	0.652	0.906	0.91	0.781
	OA = 0.833	Kappa = 0.763		

4.2 Analysis of Decomposition Results

The abundance value of decomposition inversion represents the content of the constituent abundance components of various functional types in the land use unit. The abundance of commercial land in Fig. 2(a) shows a significant tendency of "center clustering" in Luohu, Futian, and Nanshan Districts. These central areas contain many financial services or commercial streets, which have developed for a long duration and are highly mature. On the contrary, other districts such as Baoan, Longhua, and Longgang have less Commercial land use. The spatial distribution of public land in Fig. 2(b) shows a "fragmented distribution" pattern, with the build-up of public facilities being relevant to people's lives and more widespread, including some large scenic areas and sports facilities. The residential land use development pattern in Fig. 2(c) has a "multi-center clustering-sparse periphery" scenario, with high-density and extensive residential centers in Baoan, Longgang, and Longhua Districts. Besides, these areas are near the central urban spaces of each administrative district, where transportation networks and POI facilities are well laid out. Thus, the construction of traffic and urban facilities has a powerful attraction for the expansion of residential space. In contrast to residential land use, the general layout of industrial land in Fig. 2(d) is an overall trend of "clusterization of peripheral urban areas". This structure reduces the cost of land and transportation time for industrial estates. There are a vast amount of industrial parks in Bao'an, Longgang, Longhua, Pingshan, Guangming, and Nanshan districts, which are clustered to reduce costs and foster industrial synergy.

On the other hand, the spatial qualitative comparison is made between the decomposition model results and the reference data of 500 m grid-scale land units, as shown in Fig. 3. There are 8 typical representative land units in total, which correspond to 8 local areas (a to h) in Fig. 2. The subplots (a1 to h1) are 500 m classification labels of reference data, and (a2 to h2) are inversion abundance information of corresponding regions of mixed decomposition. In general, the category label of reference data can only express the dominant land use type in a certain area, but cannot express the land use abundance information of other types of information.

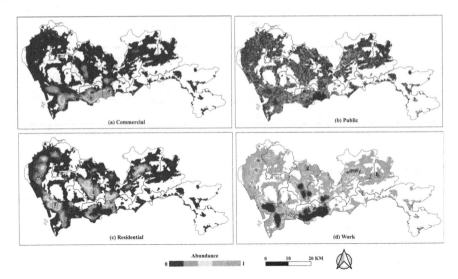

Fig. 2. Decomposition results of urban mixed-function land use

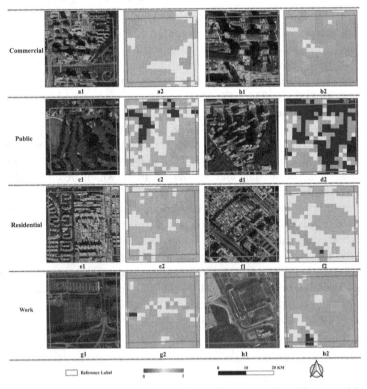

Fig. 3. Abundance information comparison of local mixed land decomposition

4.3 Quantitative Measurement for Mixed Land Use

The mixed structure of multi-functional elements inside the Shannon mixed diversity index measuring unit was constructed, as shown in Fig. 4. On the whole, the mixed degree of land use in the central urban area is higher than that in the outer marginal area. At the same time, the mixed degree of land use in Baoan, Nanshan, Longhua, Futian, and Luohu is higher, and there are several areas with a high level of mixed functional land use, with SHDI values ranging from 0.9 to 1.2 and 1.2 to 1.5. Obviously, in the periphery of the city, SHDI is in the range of 0 to 0.3, and the mixing level is low. Combined with the results of mixing decomposition it can be seen that the outer layer is mostly industrial parks and factories. At the same time, the industrial land types in Longgang District, Yantian District, Pingshan District, Guangming District, and Dapeng New District cover a wide area and have a high abundance value, and there is a lot of low-efficiency land use of single industrial use.

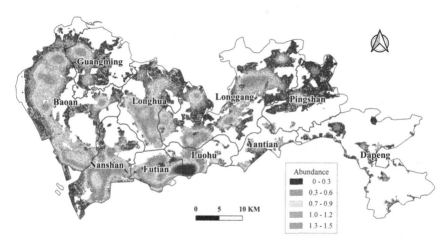

Fig. 4. The calculation results of the Shannon Diversity Indicator

5 Conclusion

In summary, based on the actual requirement of fine quantitative estimation of urban mixed land-use, this paper proposes a framework for the decomposition of urban mixed functional elements by coupling multiple data sources and combining spectral decomposition strategies, providing new scientific thinking for urban mixed research. On the one hand, the framework extends the application field of traditional spectral decomposition methods, expands the spectral bands by introducing POI and OSM data, plays a complementary role to the image data, and solves the limitation of insufficient semantic information of social functions reflected by remote sensing images.

On the other hand, it can mine the mixed situation and detailed abundance information of multifunctional urban land at a finer spatial scale, which enables decision-makers to regulate the stock land use more accurately and provide a reference basis for planning work. At the same time, the POI data and OSM road data with the advantages of easy access, easy processing, and high spatial resolution, and also use Landsat8 remote sensing images, and the fusion between them is more fine and accurate for spatial structure.

Acknowledgement. This work is supported by The National Natural Science Foundation of China (No. 41901332).

References

1. Grant, J.: Mixed use in theory and practice: Canadian experience with implementing a planning principle. J. Am. Plann. Assoc. **68**, 71–84 (2002)
2. Niu, N., et al.: Integrating multi-source big data to infer building functions. Int. J. Geogr. Inf. Sci. **31**, 1871–1890 (2017)
3. Yao, Y., et al.: Mapping fine-scale population distributions at the building level by integrating multisource geospatial big data. Int. J. Geogr. Inf. Sci. **31**, 1220–1244 (2017)
4. Zhang, J., Li, X., Yao, Y., Hong, Y., He, J., Jiang, Z.: The Traj2Vec model to quantify residents' spatial trajectories and estimate the proportions of urban land-use types. Int. J. Geogr. Inf. Sci. **35**, 193–211 (2020)
5. Zhang, Y., Li, Q., Tu, W., Mai, K., Yao, Y., Chen, Y.: Functional urban land use recognition integrating multi-source geospatial data and cross-correlations. Comput. Environ. Urban Syst. **78**, 101374 (2019)
6. Liu, X., et al.: Characterizing mixed-use buildings based on multi-source big data. Int. J. Geogr. Inf. Sci. **32**, 738–756 (2018)
7. Liang, X., Guan, Q., Clarke, K.C., Chen, G., Guo, S., Yao, Y.: Mixed-cell cellular automata: a new approach for simulating the spatio-temporal dynamics of mixed land use structures. Landsc. Urban Plan. **205**, 103960 (2021)
8. He, J., et al.: Accurate estimation of the proportion of mixed land use at the street-block level by integrating high spatial resolution images and geospatial big data. IEEE Trans. Geosci. Remote Sens. **59**(8), 6357–6370 (2020)
9. Liu, Y., et al.: Social sensing: a new approach to understanding our socioeconomic environments. Ann. Assoc. Am. Geogr. **105**, 512–530 (2015)
10. Xing, H., Meng, Y., Shi, Y.: A dynamic human activity-driven model for mixed land use evaluation using social media data. Trans. GIS **22**, 1130–1151 (2018). https://doi.org/10.1111/tgis.12447
11. Bioucas-Dias, J.M., et al.: Hyperspectral unmixing overview: geometrical, statistical, and sparse regression-based approaches. IEEE J. Select. Top. Appl. Earth Observ. Remote Sens. **5**, 354–379 (2012)
12. Gao, S., Janowicz, K., Couclelis, H.: Extracting urban functional regions from points of interest and human activities on location-based social networks. Trans. GIS **21**, 446–467 (2017)
13. Mohanaiah, P., Sathyanarayana, P., GuruKumar, L.: Image texture feature extraction using GLCM approach. Int. J. Sci. Res. Publ. **3**, 1 (2013)
14. Xie, Z., Yan, J.: Kernel Density Estimation of traffic accidents in a network space. Comput. Environ. Urban Syst. **32**, 396–406 (2008)
15. Plaza, A., Martín, G., Plaza, J., Zortea, M., Sánchez, S.: Recent developments in endmember extraction and spectral unmixing. In: Prasad, S., Bruce, L.M., Chanussot, J. (eds.) Optical Remote Sensing: Advances in Signal Processing and Exploitation Techniques, pp. 235–267. Springer, Heidelberg (2011). https://doi.org/10.1007/978-3-642-14212-3_12

A Local Spatial Kriging Applied to the PM2.5 Concentration Estimation

Shiqi Yao[1] and Bo Huang[1,2,3(✉)]

[1] Department of Geography and Resource Management, The Chinese University of Hong Kong,
Hong Kong SAR, China
bohuang@cuhk.edu.hk
[2] Institute of Space and Earth Information Science, The Chinese University of Hong Kong,
Hong Kong SAR, China
[3] Shenzhen Research Institute, The Chinese University of Hong Kong, Hong Kong SAR, China

Abstract. Air pollution in China has aroused special concerns from the general public in recent years. Accurately estimating air pollutant concentrations, especially fine particulate matter (i.e., PM2.5), is of importance to understand the distribution patterns of pollutants and then facilitate the control of their emissions and the reduction of public exposure. While estimating ambient PM2.5 concentrations based on monitoring station measurements is inherently complex, geostatistical spatial interpolation could provide a potential solution. However, standard Kriging may fail in yielding an accurate estimate as it assumes a universally homogeneous semi-variogram and lacks in considering the non-stationary spatial process that arises from anisotropy and non-randomized point pattern. This study proposes an algorithm named Point Pattern Local Anisotropy Kriging (PPLAKriging) that can derive a locally adaptive semi-variogram. Specifically, this algorithm determines an optimum neighboring search range in terms of the local distribution pattern of surrounding points followed by a local anisotropy detection. A geographical coordinates transformation is then performed for ease of calculation. To validate the proposed model in interpolating PM2.5 concentration, a regional experiment is conducted. The results show that there are substantial benefits in considering local features under different neighborhood environmental conditions. In the experiment, PPLAKriging proves to be superior to the standard ordinary Kriging, root-mean-squared-error decrease by 33.9%.

Keywords: Kriging · Anisotropy · Point pattern · Fine particulate matter

1 Introduction

Fine particulate matter, a complex small size of particles suspends in the air with diameter less than 2.5 μm, is convinced to be harmful to public health. There is strong epidemiological evidence from public health realm that air pollution especially PM2.5 has a high correlation with the disease of respiration morbidity and mortality in these years [1, 2]. However, PM2.5 concentrations in those researches are measured from a finite number of ground-based monitoring stations, which makes them hard to take fine-grained

© Springer Nature Switzerland AG 2021
G. Pan et al. (Eds.): SpatialDI 2021, LNCS 12753, pp. 205–221, 2021.
https://doi.org/10.1007/978-3-030-85462-1_19

variations of air pollution into consideration. This limitation of sparse distribution of monitoring stations leads to an urgent need for PM2.5 concentration interpolation.

In these years, extensive studies have been conducted on using interpolation approaches to estimate PM2.5 concentrations. Among them, studies can be generally grouped into two classes: algorithms in the first class are supported by multiple data sources, and the second class tends to explore features within data using statistics approaches. Studies in the first class typically integrates data from multiple sources (e.g., satellite imageries, meteorological conditions, or land-use type), or construct regression model for filling missing data [3–5]. While these algorithms have ability in estimating PM2.5 concentrations with high accuracy, they may fail in applying to real-time applications because of the low and different update frequency (e.g., half-day update frequency for satellite imagery; one-year update frequency for land-use data) of data sources. By exploring features within data to construct appropriate models [6, 7] for describing variations in data, algorithm in the second class are more suitable in real-world experiments compared to multivariate dataset-based approaches. These features, especially local features, are suggested being important in modelling variations of a given study area [8]. Therefore, it is beneficial to allow local features into interpolation approach.

Kriging, an interpolation approach, estimates value for an unknown place based a fitted variogram which can account for spatial variations with the assumption of isotropy and homogeneity. However, these assumptions are always broke because the geographical spatial patterns are naturally heterogeneous. In addition, there are many studies reported that environmental phenomena (e.g., PM2.5 concentration) has stronger correlation within a small scale; corresponding variation patterns are different under variety spatial scales and different directions [9–11]. Consequently, modelling anisotropic and spatially heterogenous are essential for describing spatially variations, and can provide higher interpolation accuracy. To specify these local characteristics, a variety of methods have emerged associated with considering spatial heterogeneous and anisotropic pattern [12, 13]. Machua-Mory et al. [12] proposed a Kriging approach for grade modelling. They constructed a local Kriging based on detected local anisotropy angles which is directed measured from the distribution of grades. Although this study gave a practical view for use in mineral resource estimation and presented high efficiency in case study, the assumption that anisotropy is always presence introduces a series of steps for counting anisotropy, and thus increase time complexity. Kracht and Gerboles [14] in their study derived a spatial representativeness-based Kriging (SR-Kriging), emphasized the importance of study extent for concluding air pollution variations. This SR-Kriging made significant improvements on modelling air pollution by considering the influences of study range, the lack of directional anisotropy is a major limitation to apply on multiple environmental variable estimations especially the air quality because the air pollution normally controlled by prevailing environmental conditions (e.g., wind). A most similar work to us was presented in [15] which relates local features, i.e., spatial scales, and directional anisotropy, as two essential variance structures in the data to describe the complex spatial evolution of hydroclimatic process. However, this study first posited that spatial scale has a greater influence against anisotropy which might not be suitable for modelling spatial process variance of air quality. Furthermore, a "moving window" strategy used in this study for detecting a suitable neighborhood largely increase the computation complexity. Consequently, an algorithm specifically designed for real-time air pollution estimation is needed to be developed.

In this paper, we propose an improved Kriging based on detected local characteristics, namely Point Pattern Local Anisotropy Kriging (PPLAKriging), to interpolate PM2.5 concentrations by constructing a locally adaptive semi-variogram. In this study, we present how the impacts of local features on interpreting spatial variances, and show the use of features for fitting semi-variogram. We first explore the spatial representativeness (i.e., point patten) in data with a suitable searching neighborhood range is given, detect directional anisotropy (directions and magnitude) from neighbors, and thus fit a local adaptive semi-variogram. To validate the proposed model, an experiment on a fine particulate matter dataset of Beijing over one-year is conducted as benchmarked against estimation with ordinary Kriging (OK). In addition, a sub-component of PPLAKriging (i.e., Point Pattern Kriging (PPKriging)) that only consider point pattern is also assessed.

The paper is organized as follows: Sect. 2 first describes the proposed PPLAKriging in detail, including the strategy for point pattern analysis and local anisotropy detection. Section3 presents the proposed model applied on air quality estimation; the model performance and experimental result are also shown in this section. The conclusion is presented in Sect. 4.

2   Method

2.1   Ordinary Kriging

We consider a spatial process $Z = \{z(s)|s \in D\}$ of n observation points, $s_i, i = \{1, 2, \ldots, n\}$ on the domain D, represented by two-dimensional coordinate systems. The Ordinary Kriging [16] predicator for a non-observed location s_o is generally as below:

$$\hat{Z}(s_0) = \sum_{i=1}^{n} \lambda_i Z(s_i) \tag{1}$$

where $\hat{Z}(s_0)$ is the predicted value, $Z(s_i)$ denotes observed value of n spatial points in the domain D. λ_i are the set of optimum Kriging weights, that can be solved by typical Kriging constraints when semi-variances are given. A typical semi-variance is calculated.

$$\gamma(s_i - s_j) = \frac{1}{2} Var(Z(s_i) - Z(s_j)) \tag{2}$$

With semi-variance γ are calculated from observed data, typical spatial variogram functions (e.g., spherical, power, Gauss and etc.) are selected to fit the semi-variogram surface which is formed by averaging semi-variances. To generalize the fitting power, this study employs the spherical model with the typical formation is:

$$\gamma(|h|) = \begin{cases} m\left(1.5\dfrac{|h|}{a} - 0.5\left(\dfrac{|h|}{a}\right)^3\right) & \text{if } |h| \leq a \\ m & \text{if } |h| \geq a \end{cases} \tag{3}$$

where m is sill, α denotes the range, and h represents the spatial distance lag.

2.2 Local Spatial Kriging

As discussed above, non-randomized point pattern (i.e., "clustered" pattern, "regular" pattern, and "dispersed" pattern) and geometric anisotropy are two major factors that directly contribute to local characteristics. To account for these local features, point pattern Kriging (PPKriging) and point pattern local anisotropy Kriging (PPLAKriging) are proposed with the schematic framework is presented in Fig. 1. First, this algorithm uses ordinary Kriging to employ the initial parameter settings (in Sect. 2.1). The distribution of neighborhood patterns of an unknown point is detected through two spatial statistics tests; the extent of neighborhood is then determined (i.e., shortened range for "clustered" pattern; extended range for "dispersed" pattern). With a proper spatial scale is determined by understanding point pattern, anisotropy is then detected through constructing directional variance structures. Last, local spatial Kriging is completed with the information of local characteristics.

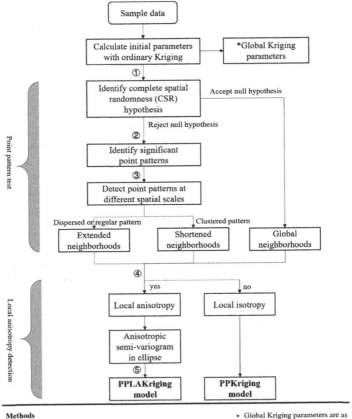

Methods

① Quadrat method to test complete spatial randomness hypothesis
② K-nearest neighbor statistics test
③ Ripley's K and L function
④ Directional anisotropy variogram
⑤ Coordination transformation

∗ Global Kriging parameters are as basic value for modification according to local characteristics.

Fig. 1. The schematic framework of the proposed PPLAKriging model.

Point Pattern Analysis

Spatial points are naturally unevenly distributed. To identify point pattern of a given study region, a sequence of three tests for point pattern distribution are applied: Quadrat method, k-nearest neighbor test, and Ripley's K function. The Quadrat method is a direct test of the complete spatial randomness (CSR) hypothesis; the latter nearest-neighbor method detects whether the study extent is being clustering or dispersion; the last method presents how the point pattern is varied under different scales.

To identify whether the whole study area is a CSR hypothesis, a first-order property of a given spatial process is evaluated with that in CSR. Typically, the CSR is constructed by building Poisson process, induced by:

$$\Pr\left(N(s) = k | \lambda\right) = \frac{(\lambda ds)^k}{k!} e^{-\lambda ds}, k = 0, 1, 2 \ldots \tag{4}$$

Here, λ denotes the expected first-order intensity, s represents a study domain. k is a finite order. ds denotes the area of study region. Following the Quadrat method, distribution of points is evaluated in finite separated integer rectangles that partitioned from study area. These partitioned rectangles must have same area. Then, the null hypothesis is defined as,

$$H_0 = Points\ exhibit\ complete\ spatial\ randomness (CSR) \tag{5}$$

against alternative hypothesis,

$$H_1 = Points\ are\ spatially\ clustered,\ dispered,\ or\ reguler\ distributed \tag{6}$$

The spatial domain is regarded to be randomly distributed when the null hypothesis is accepted, and vice versa. If a spatial domain is identified as a CSR process, the global Kriging parameters are applied for the interpolation.

For study area, evaluated as non-randomized point pattern, is then determined to be clustering or dispersion by k-nearest neighbor test. The k-nearest neighbor test is based on an evaluation of intensity,

$$\lambda(s) = \lim_{ds \to 0} \frac{E(N(ds))}{|ds|} \tag{7}$$

where $\lambda(s)$ represents the first-order intensity, ds denotes the extent of region with the s is the location. By constructing a distribution function on intensity of different distance from a point to the nearest other point, a comparison over the distribution function of CSR is followed. With the measured Kth nearest neighbor distribution, point pattern (a significant clustering or dispersed) of a point's neighborhood can be identified.

The above two tests are effectiveness in detecting significant non-stationary process, whereas invalid for discriminating point patterns that performance limited heterogeneity. This limited variations in point pattern can be identified by measuring the variation under different spatial scales. To describe inner variations of a given Kth neighbors at many distances, Ripley's K function is used. For a searched neighbor of an unknown location, Ripley's K gives statistics result by quantifying how the deviates is when comparing the Kth neighbors against a designed CSR process in multiple scales. To avoid biases arises from edge effects when calculating K function, an edge-corrected Ripley's K function [17] is given as below:

$$K(h) = \left(\frac{A}{n^2}\right) \times \sum_i^n \sum_j^n \left(\frac{I_h(d_{ij})}{w_{ij}}\right) \tag{8}$$

Here, $\left(\frac{A}{n^2}\right)$ is the density of points, and A is the area of the study region, N is the observed number of points. d_{ij} is the distance between the observed point i and j with $I_h(\cdot)$ is an indicator function. When the indicator function has a value of 1, the distance between two observed points is smaller than a given interval distance, and 0 otherwise. w_{ij}, a weight function for accounting edge-correction, gives a conditional probability that neighborhoods of point i or j falls in the study area. It should be noted that, the area of local region in this study is set as the area of 20 neighborhoods for each unobserved point. As above, this K statistics summarize how the point pattern in neighborhoods are varied under different distances.

While the Ripley's K function gives an approximate description of the distribution pattern, the discrimination between the expected value and the calculated value is still illegibility. Therefore, a transformation of Ripley's K function (i.e., Ripley's L function) is applied:

$$L(h) = \sqrt{\frac{K(h)}{\pi}} - h \tag{9}$$

By transforming the (8) to (9), the L function can discern between the value of study area and the expected value of CSR, the latter is constructed by $K(h) = \pi h^2$. Therefore, the pairwise relations for discriminating agglomeration distribution and dispersion distribution can be given the following simpler form in terms of L function,

$$L(h) > 0, \; clustering \; at \; scale \; h \tag{10}$$

$$L(h) < 0, \; dispersed \; at \; scale \; h \tag{11}$$

Geometric Anisotropy
In global Kriging, isotropy is always assumed in fitting variogram which fails to describe environmental phenomena that has prominent local characteristics. Take air pollution estimation as an application, PM2.5 concentration always exhibits directionality when wind is prevailing along one direction. This indicates the directional features must be considered when the variable is strongly influenced by neighboring environmental conditions. Furthermore, neighboring points that are located within the prevailing direction have larger influences on unknown points compared to points at other directions. An illustration is given in Fig. 2.

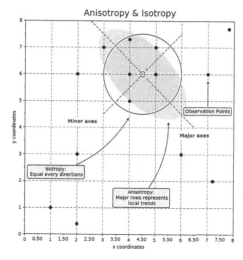

Fig. 2. The anisotropy and isotropy pattern around an unknown point.

To incorporate directionality features in the ordinary Kriging, anisotropy variograms are calculated for eight directions, i.e., each degree direction from 0 to 180° with 22.5° tolerance. With the calculated directional variograms, the preferable direction (i.e., minimum variance direction) in the neighborhood can be identified in terms of magnitudes. Note that in this study, the maximum variance direction is assumed to be perpendicular to minimum variance direction for ease of calculation. For an identified anisotropic area, we term the preferable direction is major direction and the perpendicular direction is minor direction; range in these two directions are denoted as r_{major} and r_{minor}. Accordingly, semi-variogram of major direction and minor direction are fitted in the spherical model:

$$\gamma_{major}(h_{major}) = m\left(\frac{3}{2}\frac{h_{major}}{a_{major}} - \frac{1}{2}\left(\frac{h_{major}}{a_{major}}\right)^3\right) \tag{12}$$

$$\gamma_{minor}(h_{minor}) = m\left(\frac{3}{2}\frac{h_{minor}}{a_{minor}} - \frac{1}{2}\left(\frac{h_{minor}}{a_{minor}}\right)^3\right) \tag{13}$$

Here, m is sill and same in two directions. When anisotropic variograms are calculated, the shape of searching neighborhood is in ellipse. To fasten the computation process caused by variogram calculation in two direction, a coordinate transformation is applied based on the polar coordinate system. Let the rotated angle of major direction as θ, and the axes with rotated angle θ to be the principal axes. The transformation based on two-dimensional coordination is in sequence as below:

$$\begin{pmatrix} x' \\ y' \end{pmatrix} = \begin{pmatrix} \cos\theta & \sin\theta \\ -\sin\theta & \cos\theta \end{pmatrix}\begin{pmatrix} x \\ y \end{pmatrix} \tag{14}$$

$$\begin{pmatrix} x^* \\ y^* \end{pmatrix} = \begin{pmatrix} 1 & 0 \\ 0 & \lambda \end{pmatrix}\begin{pmatrix} x' \\ y' \end{pmatrix} \tag{15}$$

$$\begin{cases} x^* = xcos(\theta) + ysin(\theta) \\ y^* = -x\lambda \sin(\theta) + y\lambda \cos(\theta) \end{cases} \tag{16}$$

With the transformed coordinates and fitted variogram, the PPLAKriging is adaptive for locally anisotropic conditions.

2.3 Performance Evaluation

To assess interpolation accuracy of the proposed PPLAKriging and PPKriging, four testing criterions are used: Root mean squared error (RMSE), standard deviation (std), spearman correlation R^2, and covariance ρ. The RMSE depicts the precision of fit and ground truth value and a lower RMSE value indicates a higher accuracy,

$$RMSE = \sqrt{\frac{1}{n}\sum_{i=1}^{N}\left[z(x_i) - \hat{z}(x_i)\right]} \tag{17}$$

where the $\hat{z}(x_i)$ is the prediction value in an unsampled point. N is the number of prediction point. The standard deviation describes the variance of error, and is given by,

$$std = \sqrt{\frac{\sum_{i=1}^{n}(x_i - \bar{x})^2}{n - 1}} \tag{18}$$

A greater std value implies a greater variation within error, gives a prove that the proposed model is not robust to the application compared to other models.

3   Experimental Result and Discussion

3.1 Study Area and Data

To examine the applicability of the PPLAKriging and PPKriging, a case study was implemented using PM2.5 concentrations value observed from 1st January to 31st December, spanning a total of 365 days, in Beijing, China. There are 35 environmental monitoring stations in Beijing, most of them are spaced out in central and southern parts of Beijing. As portrayed in Fig. 3, the study area is approximately 10,600 km^2, with monitoring stations spread out over a distance of some 150 km. PM2.5 concentrations are measured in an hourly time scale, data recorded in micrograms per cubic meter, $\mu g/m^3$. The dataset used in the validation process is the daily PM2.5 concentration for one year (from 1st Jan 2018 to 31st Dec 2018). The hourly PM2.5 concentrations were obtained from the Beijing Municipal Environmental Monitoring Center (http://zx.bjmemc.com.cn/). The daily PM2.5 concentration was the daily average of hourly PM2.5 concentrations.

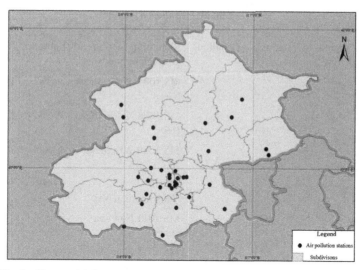

Fig. 3. The spatial distribution of 35 air quality monitoring stations in Beijing

3.2 Performance of Detecting Local Spatial Features

To examine the applicability of the proposed local spatial Kriging (i.e., PPLAKriging and PPKriging) for interpolating PM2.5 concentrations, the dataset was divided into two parts: fitting dataset and validation dataset. Four monitoring stations are randomly chosen from three-point patterns (clustered, regular, and dispersed pattern) were used for the Hold-out validation (HV); the remaining 31 monitoring stations were randomly chosen as the fitting dataset. A description of these monitoring stations is listed in Table 1.

Table 1. Description of names and spatial coordinates of the monitoring stations.

Monitoring stations to be predicted	Latitude	Longitude	Monitoring stations to be neighborhoods	Latitude	Longitude
MS.1012 Nongzhanguan	39.937	116.461	MS.1001 Bei Bu Xin Qu	4.09	116.174
MS.1018 Yi Zhuang	39.795	116.506	MS.1002 Zhi Wu Yuan	40.002	116.207
MS.1025 Men Tou Gou	39.937	116.106	MS.1003 Gu Cheng	39.914	116.814
MS.1030 Ding Ling	40.292	116.22	MS.1004 Yuan Gang	39.824	116.146
			MS.1005 Fang Shan	39.742	116.136
			MS.1006 Wan Liu	39.987	116.287

(*continued*)

Table 1. (*continued*)

Monitoring stations to be predicted	Latitude	Longitude	Monitoring stations to be neighborhoods	Latitude	Longitude
			MS.1007 Ao Ti Zhong Xin	39.982	116.397
			MS.1008 Xi Zhi Men Bei	39.964	116.349
			MS.1009 Wan Shou Xi	39.878	116.352
			MS.1010 Yongdingmen Nei	39.876	116.394
			MS.1011 Nan Sna Huan	39.856	116.368
			MS.1013 Dong Si	39.929	116.417
			MS.1014 Dong Si Huan	39.939	116.483
			MS.1015 Guan Yuan	39.929	116.399
			MS.1016 Feng Tai Park	39.863	116.279
			MS.1017 Da Xing	39.718	116.404
			MS.1019 Tong Zhou	39.886	116.663
			MS.1020 Tian Tan	39.889	116.395
			MS.1021 Qian Men	39.899	116.395
			MS.1023 Shun Yi	40.127	116.655
			MS.1024 Chang Ping	40.217	116.655
			MS.1026 Ping Gu	40.143	117.1
			MS.1027 Huai Rou	40.328	116.628
			MS.1028 Mi Yun	40.37	116.832
			MS.1029 Yan Qin	40.453	115.971
			MS.1031 Ba Da Lin	40.365	115.988
			MS.1032 Mi Yun Shui Ku	40.499	116.911
			MS.1033 Dong Gao Cun	40.1	117.12
			MS.1034 Yong Le Dian	39.712	116.783
			MS.1035 Yu Fa	39.52	116.3
			MS.1036 Liu Li He	39.58	116

Result of Point Distribution Pattern

Using the air quality dataset, the k-nearest neighbor (k-NN) statistics was first carried out to explore the point distribution pattern and the results are reported in Table 2. The overall results indicate that the k-NN test can describe point patterns by accounting the intensity value under different k values. Generally, the intensity of 30 neighbors of four regions are similar (around 1.798) which are about the average condition of global study area. More specifically, MS.1012 and MS.1018 have the highest intensities per unit area (greater than 20 when k smaller than 11) and share similar trends (a high value with neighbors less than 10; significant decreased intensities after $k = 10$, -25.77 and -25.21 respectively), whereas the neighborhood conditions of MS.1025 and MS.1030 are opposite. The larger departures between intensity of per unit area and the value of global region, the higher the count of points, and the lower the probability of the local region to be sparsely. Therefore, neighborhoods of MS.1012 and MS.1018 are easily identified as clustering patterns.

Table 2. The intensity[a] of four unobserved monitoring stations at different k values with k-nearest neighbor (k-NN) test.

SID	$K = 5$	$K = 10$	$K = 20$	$K = 30$
MS.1012	27.8	29.8	4.83	1.67
MS.1018	47.03	30.9	5.69	1.885
MS.1025	8.284	7.60	6.46	1.98
MS.1030	2.379	2.39	3.61	1.66

[a]The intensity here is multiply by e06 for an easy of representation.

While the reported k-NN statistics above can discriminate a significant clustered pattern, the difference between regular and dispersed pattern is still illegible. To further explore the variation within neighborhood of each unobserved point in detail, results of Ripley's L function are listed in Table 3. By comparing the variation between observed L value [L_o] and expected L value [L_e] that derived from CSR, differences indicates that neighborhoods of these points are extremely heterogeneity. It can be found that, for MS.1030, the L_o are negative at most distance intervals (except the 1st interval), which indicates the neighborhood of this station are tended to be dispersion. This obviously dispersion pattern of MS.1030 corresponding well with the geographical distribution, that located in western Beijing and has a close distance with Western Hills of Beijing. Contrary, other three monitoring stations all have positive L_o value at most distance intervals. Interesting, the MS.1025 performs a slightly increasing trend from 1st interval to the 5th interval (from 22.21 to 86.33), followed by a gradually drop (decreasing from 86.33 in 5th to -1.38 in 8th interval). Comparing with points that are clustered pattern (i.e., MS.1012 and MS.1018), this gradual change and a persistency lower value of L_o of MS.1025 indicates that this point can be regarded as a regular distributed pattern.

Table 3. The observed and expected L value for these four unobserved monitoring stations in nine distance scales.

SID		1	2	3	4	5	6	7	8	9
MS.1012	L_o	2.70	70.32	105.56	111.24	109.71	97.31	69.43	47.64	22.87
	L_e	0.03	0.06	0.09	0.12	0.15	0.18	0.20	0.23	0.26
MS.1018	L_o	97.04	120.6	114.67	106.24	93.06	39.64	7.93	−17.37	−61.35
	L_e	0.03	0.05	0.08	0.11	0.13	0.16	0.19	0.21	0.24
MS.1025	L_o	22.21	32.87	58.00	67.48	86.33	57.63	32.66	−1.38	7.70
	L_e	0.03	0.05	0.08	0.10	0.13	0.16	0.18	0.21	0.24
MS.1030	L_o	1.62	−24.58	−57.65	−82.07	−176.93	−327.90	−503.97	−675.81	−853.40
	L_e	0.06	0.12	0.18	0.24	0.30	0.36	0.42	0.48	0.54

Result of Local Geometric Anisotropy

Local anisotropy in the PM2.5 spatial process arises from neighboring environmental condition, dynamically changed with meteorological condition and has impacts on modelling the spatial process. To elaborate the locally detected anisotropy, directional semi-variograms of selected four monitoring stations are presented in Fig. 4 (take Dec 1st in 2018 as a case); the corresponding anisotropy setting is listed in Table 4. Generally, the variographic result shows a preferable direction (about 135°) in this spatial process. There are no significant variations in minimum variation directions (anisotropy ratio is around 1.5–2.5; angle is about 135°) in these four monitoring stations. A possible explanation for this consistency in anisotropy is that mesoscale environmental conditions are generally dominant than microscale environmental. While these four regions

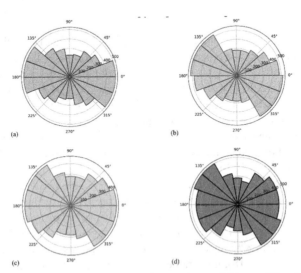

Fig. 4. The Local Anisotropy generated from four stations neighbors of Dec 1st, MS 1012; (b) MS 1018; (c) MS 1025; (d) MS 1030.

hold similar local anisotropic characteristics, variations among them are still significant. For example, MS.1030, located in a dispersed region, has largest variances in directional variograms in eight directions (semi-variance is least in about 140°; larger in 45°; largest in 60°), and the magnitude performance the stepped structure. Comparatively, MS.1025 is more homogeneity, with the magnitude difference of major and minor axes is smaller than 200. This further proves that a directional anisotropy detection for each unknown station is needed in order to describe locally variation precisely. Interestingly, it is also noticed that MS.1012 and MS.1018 has similar features in anisotropy angles and magnitude.

Table 4. Final anisotropy setting for four local stations on Dec 1st, 2018.

Parameter	Local region			
	MS.1012	MS.1018	MS.1025	MS.1030
Angle	175	149	10	140
Major	550	520	450	500
Minor	250	280	280	250
Anisotropy ratio	2.2	1.857	1.607	2

3.3 Performance of Local Spatial Kriging

Using the dataset, spatial Kriging-based models (i.e., ordinary Kriging, PPKriging and PPLAKriging) were tested, and the HV results are reported in Table 5. In general, the value of correlation coefficient ρ indicates that three models all can provide good performance in interpolation. It should be noted that the performance of local dynamic framework has increased from 0.962 in global stationary framework (i.e., OK) of ρ to 0.963 in PPKriging, and 0.978 in PPLAKriging. The increased value in covariance (from 1824 for OK to 1830 for PPKriging; from 1824 for OK to 1841 for PPLAKriging) also indicate that there is a significant improvement in PPLAKriging. As expected, PPLAKriging is the best because local anisotropy and point distribution pattern are taken into account. By comparing the standard deviation, the increased value in four testing criterions further indicate that PPLAKriging gives a better interpolation of data than other models. Interestingly, although the correlation in PPKriging shows a good fitting performance compared to OK, the increased value of standard deviation and RMSE (0.01 for standard deviation; 0.006 for RMSE). A possible reason for this is that when only considering point pattern, the Kriging may constrain to its neighborhood range. As expected, Table 5 reveals that PPLAKriging is the best with a lower RMSE value (33.9%) and lower standard deviation (25.1%). We posit that this is because PPLAKriging can handle both point distribution pattern and spatially heterogeneities. Moreover, this also indicate that the higher accuracy because of a stronger dependency of directional anisotropy than point pattern.

Table 5. Performance of PM2.5 concentration estimation using three different Kriging models.

Method	Covariance (*cov*)	Correlation (ρ)	Standard deviation (*std*)	RMSE
Global stationary framework				
Global isotropic	1825	0.962	10.783	6.588
Local dynamic framework				
PPKriging	1830	0.963	10.793	6.592
PPLAKriging	1841	0.978	8.07	4.354

Table 6. Performance of PM2.5 concentration estimation in four monitoring stations using three different Kriging model.

Method	MS	Covariance (*cov*)	Correlation (ρ)	Standard deviation (*std*)	RMSE
Global stationary framework	1012	1910	0.985	6.089	6.092
Global isotropic	1018	1892	0.968	10.352	6.768
	1025	1829	0.867	11.122	6.964
	1030	1679	0.951	12.803	8.333
Local dynamic framework	1012	1911	0.983	6.304	6.307
PPKriging	1018	1880	0.969	9.973	6.618
	1025	1856	0.967	11.042	7.069
	1030	1686	0.947	13.199	8.470
PPLAKriging	1012	1906	0.983	6.269	6.177
	1018	1878	0.969	9.971	4.398
	1025	1927	0.978	8.694	4.876
	1030	1769	0.973	9.504	5.188

As demonstrated above, the PPLAKriging shows significant improvements on OK in terms of a decreased RMSE and *std* value. The validity of the proposed model in different point patterns is another issue that we focus on. Table 6 provides a detailed comparison. Generally, the PPLAKriging model gives the least deviations in four monitoring stations. Specifically, for the clustered neighborhood, OK performs best, followed by PPLAKriging. PPKriging has the largest deviations. For the regular region (MS.1025), PPLAKriging resulted in higher accuracy and represented greater robustness with a reduced value of the deviation of 2.41. This improvement is significant in the dispersed region that PPLAKriging decreases a −3.30-error compared to OK. It can be seen that the PPLAKriging is also valid in the dispersed neighborhood (MS.1030) with an increased *cov* of 90, decreased *std* of 3.299, and decreased RMSE of 3.145, compared to the clustered region (i.e., MS.1012 and MS.1018). This further proves that the effects of local anisotropy are obvious in a dispersed region than a clustered region. This is because the spatial process in a clustered region is more homogeneous even if the variance exists in neighborhoods. We also notice that PPKriging has lower accuracies especially in dispersed or clustered pattern. This also imply that the anisotropy is a dominant factor of non-stationary spatial process in a dispersed region than point pattern.

To further explore the robustness of the proposed models in interpolating PM2.5 concentrations, we analyze the time series estimation trend in four monitoring stations. The error of daily time series estimation of PM2.5 concentration were summarized in Fig. 5. Generally, the PPLAKriging exhibits a lower estimation error compared to OK, especially in the dispersed neighborhood. Meanwhile, the improvement of the point pattern local anisotropy Kriging in a regular distribution neighborhood is also effective. Despite the local anisotropy were detected (discussed in Sect. 3.2.2), the temporal trends of estimation error in MS.1012 (clustered pattern) exhibit a not significance improvement. We can infer about the clustered pattern from this that the geometric anisotropic features are weaker within a same neighborhood range (or, neighborhood points) than that in a dispersed region; the spatial process in a clustered region tends to be more homogeneous than that in a dispersed pattern when the study extents are same in two patterns. As proved by the results in Fig. 5, the proposed PPLAKriging has the predictive power for interpolation an unknown point under a locally dynamic spatial process.

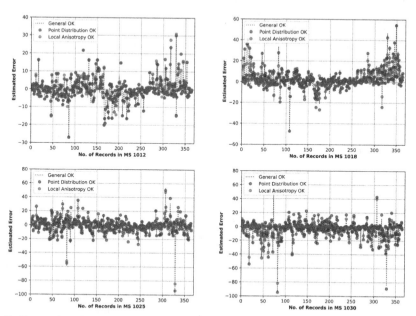

Fig. 5. Scatterplot of estimation error between estimated PM2.5 concentration against ground-truth value of four monitoring stations using OK, PPKriging, and PPLAKriging.

4   Conclusion

Previous studies demonstrated assumption of homogeneity within the neighborhood of an unknown point is unlikely to be supported, this study took a local approach that extended the ordinary Kriging for interpolating PM2.5 concentration. Our analysis reveals the local anisotropy and point patterns have significant impacts on interpolation accuracy; experimentally proved that global OK is inadequate to model spatial nonstationary; and emphasizes a dominant factor of anisotropy in the dispersed pattern.

This study extended Ordinary Kriging model to incorporate point distribution pattern and anisotropic semi-variogram (PPLAKriging) to deal the non-stationary in spatial process that arises from two local characteristics: anisotropy and non-randomized point pattern. The proposed PPLAKriging achieved the best interpolation accuracy, because it depended on the two local features simultaneously. Compared to OK, PPKriging and PPLAKriging increased the correlation from 0.962 to 0.963 and 0.978, decreased the RMSE from 6.588 to 6.592 and 4.354. The 25.15% improvement of PPLAKriging in terms of the decreased standard deviation and 33.9% value of a declined RMSE over OK imply the effectiveness of the model in modelling variogram under different neighborhoods, and the robustness in interpolating the data in an unknown place in variety meteorological conditions. This study concludes that variance in PM2.5 data varied with a complex function of point pattern and anisotropy; interpolation accuracy is extremely sensitive to local features. More specifically, the anisotropy is dominant factor in dispersed region while isotropy is a prevailing feature in a clustered region. Consequently, this study shows a effective framework for interpolation PM2.5 concentration based on neighboring conditions. In the future work, test on the macro-scale geographical region will also need to be conducted to verify the capability in different spatial scales. We will exploit to apply the concept of local features to explore whether spatio-temporal Kriging is also sensitive towards locally dynamic features.

References

1. Brookes, M., Bouganis, C.-S.: Statistical multiple light source detection. IET Compu. Vis. 1(2), 79–91 (2007)
2. Wong, N.S.et al.: PM2.5 concentration and elderly tuberculosis: analysis of spatial and temporal associations. Lancet 390, S68 (2017)
3. Zhan, Y., et al.: Satellite-based estimates of daily NO_2 exposure in China using hybrid random forest and spatiotemporal Kriging model. Environ. Sci. Technol. 52(7), 7 (2018)
4. Tang, M., Wu, X., Agrawal, P., Pongpaichet, S., JAIN, R.: Integration of diverse data sources for spatial PM2.5 data interpolation. IEEE Trans. Multimedia 19(2), 408–417 (2017)
5. Liang, F., Gao, M., Xiao, Q., Carmichael, G.R., Pan, X., Liu, Y.: Evaluation of a data fusion approach to estimate daily PM2.5 levels in North China. Environ. Res. 158, 54–60 (2017)
6. Hueglin, C., Gehrig, R., Baltensperger, U., Gysel, M., Monn, C., Vonmont, H.: Chemical characterisation of PM2.5, PM10 and coarse particles at urban, near-city and rural sites in Switzerland. Atmos. Environ. 39(4), 637–651 (2005)
7. van Donkelaar, A., et al.: Global estimates of fine particulate matter using a combined geophysical-statistical method with information from satellites, models, and monitors. Environ. Sci. Technol. 50(7), 3762–3772 (2016)

8. Cheng, G., Li, D., Zhuang, D., Wang. Y.: The influence of natural factors on the spatio-temporal distribution of Oncomelania hupensis. Acta Tropica **164**, 194–207 (2016)
9. Fotheringham, A.S., Yue, H., Li, Z.: Examining the influences of air quality in China's cities using multi-scale geographically weighted regression. Trans. GIS **23**(6),1444–1464 (2019)
10. Yan, D., Lei, Y., Shi, Y., Zhu, Q., Li, L., Zhang. Z.: Evolution of the spatiotemporal pattern of PM2.5 concentrations in China – a case study from the Beijing-Tianjin-Hebei region. Atmos. Environ. **183**, 225–233 (2018)
11. Huang, Y., Yan, Q., Zhang, C.: Spatial–temporal distribution characteristics of PM2.5 in China in 2016. J. Geovis. Spat. Anal. **2**(2), 12 (2018)
12. Machuca-Mory, D.F., Rees, H., Leuangthong, O.: Grade modelling with local anisotropy angles: a practical point of view. In: 37th Application of Computers and Operations Research in the Mineral Industry (APCOM 2015) (2015)
13. te Stroet, C.B.M., Snepvangers, J.J.J.C.: Mapping curvilinear structures with local anisotropy Kriging. Math. Geol. **37**(6), 635–649 (2005)
14. Kracht, O., Gerboles, M.: Spatial representativeness evaluation of air quality monitoring sites by point-centred variography. Int. J. Environ. Pollut. **65**, 17 (2019)
15. Romero, D., Orellana, R., Hernández-Cerda, M.E.: Multiscale spatial variographic analysis of hydroclimatic data. Theoret. Appl. Climatol. **144**(1–2), 55–66 (2021). https://doi.org/10.1007/s00704-020-03513-9
16. Cressie, N.: Spatial prediction and ordinary kriging. Math. Geol. **20**(4), 405–421 (1988)
17. Hohl, A.: Accelarting the detection of space-time patterns under non-stationary background population. Doctor of Philosophy, The University of North Carolina at Charlotte (2018)

Evaluation of Typical Public Facilities in Shanghai Urban Functional Area

Qiang Ma[1], Liangxu Wang[2(✉)], and Xin Gong[1]

[1] School of Environmental and Geographical Sciences,
Shanghai Normal University, Shanghai, China
[2] Institute of Urban Development, Shanghai Normal University, Shanghai, China
wangliangxu@shnu.edu.cn

Abstract. Public toilets, as the most typical public facilities, reflect the civilized level and management service level of the city, and it is also an important window to establish a civilized image of the city. In the existing research, more attention has been paid to the accessibility and service scope of public toilets, while ignoring the different levels of demand for public toilets in different functional areas. This article uses POI data and TF-IDF model to identify urban functional areas, and combines OSM road network data and WorldPop population data to calculate the population travel vitality index, and evaluation of public toilet services in urban functional areas. The results show that this method fully takes into account the issue of the heterogeneity of the functional areas of public toilets and the travel vitality of the population, and the spatial analysis is more accurate and has more practical value.

Keywords: TF-IDF model · Urban functional area · Travel vitality index model · Public toilets evaluation

1 Introduction

In the past 30 years, the United Nations has put forward Agenda 21, MDGs (Millennium Development Goals), Future We Desired, SDGs (Sustainable Development Goals) and other target systems. These specific targets have gradually changed from qualitative to quantitative [1]. Among them, reducing inequality between regions (SDG 10) and building inclusive, safe, and sustainable cities and communities (SDG 11) are important indicators for building regional balance and sustainable development [2–4]. Therefore, advancing the equalization of basic public services is an important task in the construction of SDGs.

In this case, the research on the spatial layout of urban public toilets has been widely concerned. Liu believes that GIS technology should be introduced in the planning to scientifically determine the service radius of public toilets and establish a public toilet service system with balanced layout and moderate quantity [5]. Tang proposed the optimal location model of multi-criteria decision analysis by optimizing the traditional location

G. Pan et al. (Eds.): SpatialDI 2021, LNCS 12753, pp. 222–235, 2021.
https://doi.org/10.1007/978-3-030-85462-1_20

model, and verified the feasibility and reproducibility of the model [6]. Abel T et al. established GIS system for Ethiopian government departments, and provided suggestions for the location of small public facilities such as public toilets, and thought that this method could enhance the scientific nature of decision-making [7]. Osumanu IK obtained the coordinates of public toilets in Wa area of Ghana by hand-held GPS, and evaluated the accessibility and utilization rate of public toilets combined with questionnaire survey [8]. Feng put forward the demand and planning scheme of public toilets in Yaodu District in the future by studying the rationality of public toilets layout in Yaodu District of Linfen City [9]. Besides, Xu studied the spatial layout of public toilets in Ma Jiang area based on the main factors of attracting crowds, such as commerce, transportation stations, tourist attractions and education [10]. Based on the resident population statistics, Xie analyzed the quantity, quality and spatial layout of public toilets in yuhu district of Xiangtan City by using the methods of population density analysis and population distribution analysis [11]. Gao graded the population density in Qinghe District of Huai 'an City from the street level, and analyzed the layout, quality and quantity of public toilets [12]. Recently, Zhao used Baidu big data to make the density distribution map of people traveling, and analyzed the rationality of the spatial layout of public toilets in Zhuhai [13].

In the existing research, the method of research on the spatial layout of basic service facilities mostly uses the population density prediction model [13], that is, the rationality of the layout of basic service facilities is measured by the attractiveness of the regional center to the population. However, the influencing factors selected by this method are too subjective, and the applicable spatial scale has certain limitations. Therefore, it is difficult to complete the evaluation of basic service facilities on a larger spatial scale.

The demand of regional public toilets is closely related to the vitality of travel. The travel vitality index model is directly related to the city functional area, population and road network density. At present, the division of urban functional areas is mainly divided into two parts: grid and road network. Considering the complexity of data grade of urban road network and the close distance between public toilets and roads, dividing urban units by OSM will inevitably delete some public toilets, resulting in certain errors. Therefore, in this paper, the grid unit is adopted in the division of urban functional areas. However, the identification of urban functional areas is mainly divided according to POI weight and quantity proportion, while POI weight is sorted by reference to public awareness, and assigned according to its individual characteristics, which is subjective. In this paper, considering the dual influence of the number and spatial distribution of POI weights, we introduce the weighting technology TF-IDF algorithm in machine learning to calculate POI weights locally and globally.

With the rapid development of big data technology and information technology, urban spatial data sets provide the possibility to evaluate the heterogeneity and rationality of the spatial layout of infrastructure service facilities on a large scale, and also provide a new perspective for the study of urban sustainable development [14].

This paper takes Shanghai as the research area and takes a typical public facility - public toilet as an example. Through the crawler technology, the data of public toilets in Shanghai area are obtained, and the urban functional area are identified by TF-IDF model [15] combined with POI data. On this basis, the travel vitality index model is constructed by integrating the WorldPop population data and OSM road network data.

The model can evaluate the public toilet service in different functional areas, and then explore its rationality from the space. The paper puts forward the path selection and countermeasures to optimize the supply capacity of public services in Shanghai, in order to provide a reference method for the planning and evaluation of public facilities.

2   Research Area and Data

2.1   Study Area

The research area of this paper is Shanghai, which contains 16 districts, among which Huangpu District, Xuhui District, Changning District, Jing'an District, Putuo District, Hongkou District and Yangpu District are regarded as the downtown area. As early as 2006, Shanghai issued the *Outline of Shanghai Public Toilet Layout Planning*, and in 2017, it issued a new edition of *Public Toilet Planning and Design Standards*, which has a more detailed standard for the setting of public toilets.

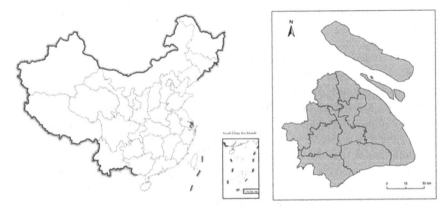

Fig. 1. Study area

2.2   Data Source

The research data used in this article includes the POI data set, that is, the 2018 AutoNavi POI data set obtained through crawler technology, which has basic information such as latitude and longitude, category, address, and name. With a total of 1,340,269 pieces of data.

Among them, there are 23,692 public toilets.

In addition, it also includes Worldpop population data and OSM road network data. Worldpop data set is based on census, survey data, social media data, night light data, land use data, etc., and estimates the weight layer of population distribution through random forest model, making a high-resolution population distribution grid map [16]. In this paper, we download the data set of China's population in World POP in 2018, and get the population in each spatial grid unit of Shanghai through mask extraction, grid transformation and spatial connection. OSM road network data adopts vector format, which is divided into different categories according to road grade and size.

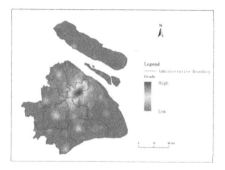

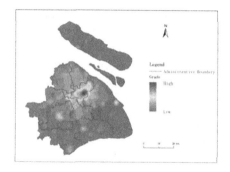

(a) All POI levels

(b) POI level of public toilets

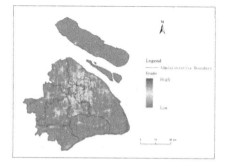

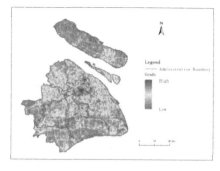

(c) Population density grade

(d) Road network density grade

Fig. 2. Data source

3   Research Process and Method

It mainly includes two parts: calculation of travel vitality index and rationality evaluation of urban functional areas. The identification of urban functional areas uses the TF-IDF model to calculate the weights and calculates with frequency density [17]. According to the weights of different functional areas, road network density, population density, and a travel vitality index model for urban functional areas are established. Finally, according to the measurement model of public toilets, the quality is evaluated and analyzed.

3.1   Calculation of Travel Vitality Index in Urban Functional Area

TF-IDF Model. The recognition of urban functional areas mainly depends on the number and weight of POIs. In real urban space, the data amount of various POIs varies greatly. For example, there are 109,806 POIs for catering services and 10,316 POIs for scenic spots. The huge difference in the number will make the small data disappear in the calculation, so the weight setting of POIs is very important [18]. Traditional POI weights refer to Zhao's public cognitive ranking of POIs [19]. How to design POI weights in different cities and different orders of magnitude becomes a difficult problem. TF-IDF is a statistical method used to estimate the importance of a word in a document set, which is often used for weighting information retrieval and data mining in machine learning, and has good classification ability [15]. The research mainly regards each spatial grid unit as a single file, and the category of POI in a single file is regarded as a word, so analyzing the category of each spatial research unit is transformed into solving the weight of each word in the file set.

$$tf_{i,j} = \frac{n_{i,j}}{\sum_k n_{k,j}} \tag{1}$$

$$idf_i = \lg \frac{|D|}{\left|\{j : t_i \in d_j\}\right|} \tag{2}$$

$$tfidf_{i,j} = tf_{i,j} \times idf_i \tag{3}$$

In the formula, i represents a certain type of POI; j represents a spatial grid unit; n_{ij} represents the number of times that class i POI appears in j space grid unit; $\sum_k n_{k,j}$ represents the sum of all POI occurrence times of spatial grid unit j; D represents the total number of spatial grid units; $\left|\{j : t_i \in d_j\}\right|$ indicates the total number of spatial grid cells containing a certain type of POI.

Travel Vitality Index Model. Different functional areas have different attraction for people's travel. The travel vitality index model uses the population and road network richness in the spatial grid unit to normalize them and then weight them. And then multiplied by the weight coefficients of different functional areas.

$$G_{xm} = \frac{G_m - G_x^{min}}{G_x^{max} - G_x^{min}} \tag{4}$$

$$C_m = G_{xm} \times Q_y (x = 1, 2; y = 1, 2, 3, \cdots, 8; m = 1, 2, 3, \cdots, n) \tag{5}$$

In the formula, $x1$ and $x2$ respectively represent the population and road network richness in the spatial grid unit; m represents the number of the spatial grid unit; G_m represents the numerical value of m cell; G_x^{max} represents the maximum value of class x, G_x^{min} represents the minimum value of class x, and G_{xm} represents the normalized value of class x of spatial grid element m; Q_y represents the weight of eight functional areas in terms of travel vitality index in turn; C_m represents the travel vitality index of the m-th spatial grid unit.

3.2 Rationality Evaluation of Public Toilet

According to the Standard for Setting Environmental Sanitation Facilities 2012 of the Ministry of Housing and Urban-Rural Development, the setting of public toilets should be combined with the urban land use type and the prosperity degree of the land use type, of which 4–11 land for public and commercial services are used per square kilometer, of which the areas with high prosperity degree should take the high limit and the areas with low prosperity degree should take the low limit; However, industrial land only needs 1–2 per square kilometer. Therefore, the planned number of public toilets is specified by combining the relevant standards and the natural discontinuity classification method of travel vitality index Table 1:

Table 1. Planned number of public toilets

Travel vitality index x	$0 < x \leq 5$	$5 < x \leq 10$	$10 < x \leq 25$	$25 < x \leq 45$	$x > 45$
Number of public toilets	1	3	5	7	10

$$H = \frac{S - P}{P} \times 100\% \tag{6}$$

Where, S is the actual number of public toilets in the grid unit; P is the planned number of public toilets in the grid unit. Through the analysis of H, we can get the relationship between the actual number of public toilets and the planned number of public toilets, and judge the service level of public toilets in this area.

The technology roadmap is shown in Figures 1, 2 and 3.

4   Calculation of Travel Activity Index in Urban Functional Areas

4.1   Functional Area Identification and Verification

The types of urban functional areas are mainly determined by POI category and quantity. Based on fishing nets, Shanghai is divided into 7275 spatial grid units of 1 km². The TF-IDF model is used to weight the number of POIs in the cell. Compared with the traditional weight assignment method, it not only considers the distribution of POIs in the whole area, but also considers their density in the spatial grid cell, which can identify the urban functional area more accurately.

The weighted values of all kinds of POI in each spatial grid unit, namely the frequency density, are calculated by using the formula, and the urban functional areas are divided according to the frequency density. Firstly, the areas without POI are screened out and divided into data-free areas; When the frequency density of a certain type of POI is greater than 50%, the attribute of this type of POI is set as the attribute of the functional area; When none of the POIs has a weight of more than 50%, the two POIs with a weight of 20%–50% are named as the joint functional areas of the two POI categories, and the rest are classified as comprehensive functional areas.

The recognition results of urban functional areas in Shanghai based on the TF-IDF model and frequency density is shown in the figure. The results show that there are 7275 urban functional areas in Shanghai, and the spatial distribution is dominated by the three

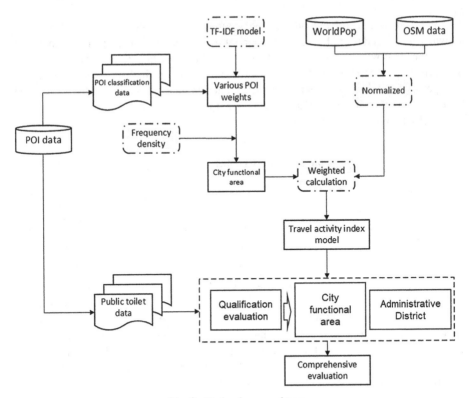

Fig. 3. Technology roadmap

major categories of commercial service, industry, public and related joint functional areas, which is highly consistent with the proportion of urban land types in Shanghai. There are 1,615 joint functional areas, which are scattered and cover a wide range, and are staggered with single functional areas; There are 4,170 single functional areas, which are mainly composed of commercial service (1,457), industrial service (1,324) and public service (750). Commercial service functional areas are mainly concentrated in Shanghai urban areas and various regional centers, which are closely related to Shanghai's highly developed commerce. Industrial functional areas account for a small proportion in Shanghai urban areas and areas close to the urban areas, and are mainly distributed in suburbs far away from the urban areas. The distribution of public functional areas is scattered, which shows that there is a certain degree of spatial balance; There are 70 mixed functional areas, mostly distributed in the center of each area; There are 1420 data-free areas, mainly lakes, farmland, beaches, water systems and other areas (Fig. 4).

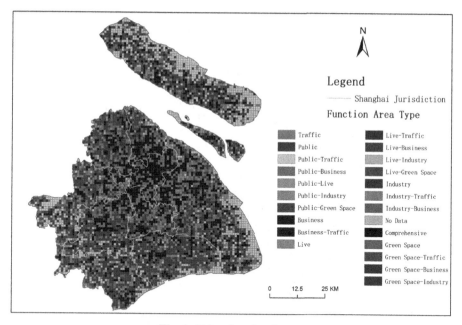

Fig. 4. Urban functional areas

4.2 Calculation of Travel Vitality Index

The service level of public toilets should be determined by the travel vitality index of the place, and the density of road network is often positively correlated with the travel vitality of the place. Such as parks, green spaces, squares, shopping streets, etc. In some areas, the density of unit road network is high, but it is mainly used as land for transportation facilities, so the travel vitality is not high, such as transportation hubs, viaducts, overpasses and so on. In this paper, a travel vitality index model is established, and through normalization and superposition analysis of population and road network density, a more detailed spatial distribution of population travel vitality in urban functional areas is obtained, as shown in the figure (Fig. 5).

It can be found by using the natural discontinuity method to visualize the trip activity index of spatial grid cells that the spatial distribution of population trip activity index has obvious characteristics of "massive aggregation polycentric development", and the spatial difference is obvious. From the spatial distribution of district level, the travel vitality index of Shanghai Central District is relatively high, with obvious scale advantage. Very prosperous and prosperous areas are mostly located in the central city, with sporadic distribution in suburban centers and suburban new towns. The most prosperous areas include the Bund, Lujiazui, Zhongshan Park, Yangpu Park and other places, showing contiguous distribution along the subway line, with the highest vitality of population travel. The more prosperous areas are mainly Qibao, Disney Park, Shanghai Jiaotong University (Minhang Campus) and other places in the urban and suburban areas of Shanghai. University towns, parks and suburban commercial centers have higher travel

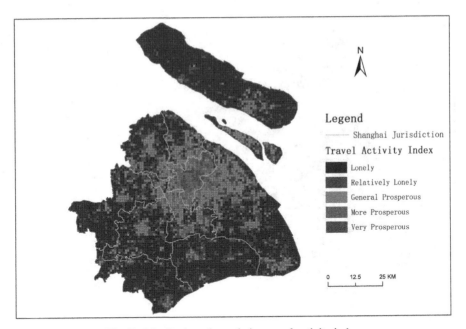

Fig. 5. Distribution of population travel activity index

vitality. The general prosperous areas are located in the obvious sub-centers of various suburbs of Shanghai, with a certain scale, such as Songjiang New City, Nanqiao New City, Qingpu New City and so on. The relatively deserted area is the edge area of suburban center and suburban new town, which has obvious annular distribution characteristics. In addition, it also includes the center of suburban street town, which has certain travel vitality. Cold and cheerless areas are the norm in the outer suburbs, such as Jinshan, Chongming, Fengxian and the south of Pudong New Area, mainly in some data-free areas and rural areas. The geographical location here is remote, with primary industry as the main factor, low population density and inconvenient transportation.

5   Evaluation of the Rationality of Public Toilets

5.1   Evaluation of Public Toilet Service

According to the relevant regulations, using the population, road network density, functional area types and the number of public toilets, the public toilet service level measurement model is used to evaluate the number of public toilets in each urban functional area. The results h were divided into five categories, which were excellent ($H \geq 0.5$), good ($0 < H \leq 0.5$), qualified ($H = 0$), unqualified ($-0.5 < H < 0$), poor ($H \leq -0.5$). Excellent areas are mainly concentrated in Shanghai downtown and Baoshan Gucun, Jiading Ma Lu, Songjiang Xinqiao, Minhang Meilong, Huacao, Qingpu Huaxin and Chuansha New Town in Pudong New Area. Among them, Shanghai is highly developed in business, and a large number of shopping malls can provide public toilets. The excellent service of public toilets in suburban towns is mainly due to the fact that the functional areas are mainly industrial areas with low travel vitality index and low requirements for public toilets. There are two excellent areas in Chuansha New Town, Pudong New Area, namely Shanghai Disney and Pudong International Airport, which are the representatives of green space and traffic functional areas. Although there is a high travel vitality index here, the public toilet facilities are perfect enough to meet people's needs. Most of the good and qualified areas are concentrated in the marginal zone of excellent areas, and the construction density of public toilets has an obvious attenuation trend from the center to the periphery, but the attenuation of travel vitality index is relatively slow, so although it is still in the standard range, the public service level is lower than that of the central area. Most of the unqualified areas are located around high-grade roads such as expressways and provincial highways, and the vitality of people's travel is relatively high. However, because driving is the main mode of travel in this area, walking is not the main mode, and it has higher mobility. Therefore, the distance between public toilets is large, and the areas with unqualified service function areas of public toilets are characterized by intermittent distribution along high-grade roads. The functional areas with poor public toilet service are mainly concentrated in the outskirts of cities, and the coverage rate of public facilities is low. For example, Nanhui New Town, as the regional center, has a large number of parks and universities around Dishui Lake, but there are few public toilets, which can hardly meet the needs of the traveling population in this area. However, there are a large number of non-built areas in Jinshan and Chongming, such as cultivated land. Although the vitality index of population travel is low, the distribution of public toilets basically only exists in the core areas such as regional town centers, so the service level of public places in the surrounding areas is poor (Fig. 6).

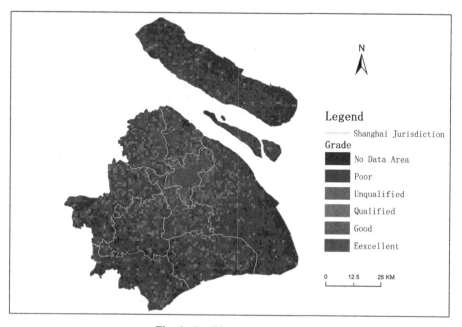

Fig. 6. Qualification evaluation

5.2 Public Toilet Service Statistics Based on Functional Areas

The setting of public toilets is largely based on the types of urban functional areas. By coupling the types of urban functional areas with the evaluation results of public toilets, it can be found that the public toilets have the highest qualification among commercial service functional areas. In addition, the qualification of joint functional areas related to commercial services such as "industry-commercial service" and "green space-commercial service" is also at a high level, which is closely related to the progress of current commercial services. Large commercial service organizations such as chain brands provide public toilet services to the outside world. Pedestrians can find the nearest public toilet by searching on the map. The level of regional economic development and the concentration of economic resources directly affect the openness of public toilet services. Surprisingly, the public functional area has the lowest qualification, which is related to the definition of public functional area, which mainly includes schools, government agencies, social organizations and other institutions and facilities. The openness of such institutions directly affects the convenience of their public services. Especially in the current epidemic situation, the relevant institutions have a low degree of openness and strict management, so it is difficult for people traveling in public functional areas to provide meaningful public resources. Therefore, the establishment of public toilets needs to be analyzed from a three-dimensional perspective to fully explore the demand (Fig. 7).

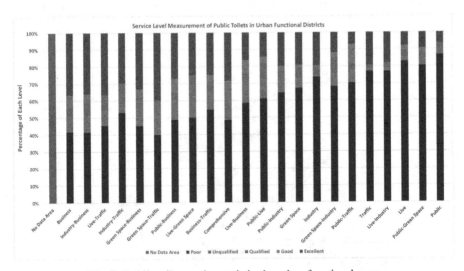

Fig. 7. Public toilet service statistics based on functional areas

5.3 Statistics of Public Toilet Services Based on Administrative Regions

Data-free areas are often urban non-built areas. Therefore, from the point of view of urban built-up areas, that is, urban functional areas within the jurisdiction, statistics are made on the eligibility of public toilets in various districts of Shanghai, and it is found that the distribution of the eligibility shows polarization effect, with a high proportion of "poor" and "excellent". Generally speaking, there are a large number of functional areas in the suburbs, showing a high value, and only the two indicators in the urban areas are at a low value, which indicates that there are a large number of gaps in public toilet services in the suburbs of Shanghai, and the unqualified level of public toilet services is relatively high, while the qualified degree of public toilets in the urban areas is obviously improved, which is mainly due to the more developed economy in the urban areas and the better construction of public service facilities. However, the ratio of "qualified" and "good" in the middle of the bar chart generally shows a low value, which indicates that the construction of public toilets is unreasonable to a certain extent. The appearance of polarization shows that the balance between supply and actual demand is difficult to grasp, and the planning and construction of urban public facilities need to be further explored. From the perspective of "excellent" areas, the excellent rate of urban areas, Baoshan District, Jiading District and Minhang District exceeds 50%, which has a great relationship with their location in the suburbs of the city. Urban expansion is often spread from the inside out. Therefore, the construction level of basic public service facilities in the suburbs is relatively high, and the outer suburbs are difficult to be radiated by urban resources and cannot enjoy the overflow of urban resources, so it is difficult to improve their public toilet service level, such as Chongming (Fig. 8).

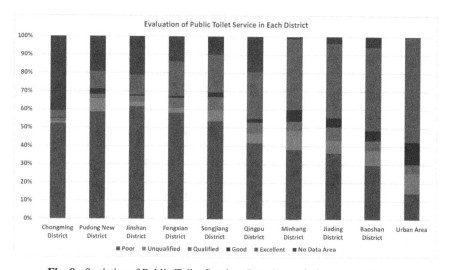

Fig. 8. Statistics of Public Toilet Services Based on Administrative Regions

6   Conclusion and Discussion

In this paper, combined with the related planning standards of China and Shanghai, the service capacity of public toilets in Shanghai is measured by using multi-source data, including service scope, population coverage and service effect. It also evaluates and categorizes from the streets and spatial grid units. The study found that there is a significant difference in public toilet service in Shanghai from the central city to the peripheral suburbs. From downtown Shanghai to the southwest, southeast, and Chongming Island, there is a significant decline. However, the decaying capacity of public toilet service from the center of Shanghai to the northwest area is not obvious, indicating the good service capacity of the continuum.

However, at the same time, we also found that WorldPop population data cannot accurately reflect the temporal and spatial distribution of population travel vitality, and the big data of POI cannot explain the specific operation of urban facilities. Therefore, how to improve the establishment standards of the travel vitality index model and the qualitative analysis of public facilities using other data set (such as social media data) will be the focus of the next step of research.

References

1. Wang, G.H., Liu, Y.J, Wang, H.B.: Succession of the world's sustainable development goals in the past 30 years. J. Bull. Chin. Acad. Sci. **30**(05), 586–592 (2015)
2. Zhang, J., Wang, S., Zhao, W.W., et al.: Progress in the study of the relationship between sustainable development goals. J. Acta. Ecol. Sinica. **39**(22), 8327–8337 (2019)
3. Zhu, J., Sun, X.Z., He, Z.: Research on China's sustainable development evaluation indicators under the SDGs framework. J. China Popul. Res. Environ. **28**(12), 9–18 (2018)
4. Zhang, J.Z., Wang, S., Zhao, W.W., et al.: Progress in the study of the relationship between sustainable development goals. Acta. Ecol. Sinica. **39**(22), 8327–8337 (2019)
5. Liu, A.S., Zhang, W., Liu, M.: A preliminary study on the spatial layout planning and digital application of public toilets——taking the spatial layout planning of public toilets in downtown Changzhou as an example. J. Urban Dev. Res. (S1), 188–191+225 (2008)
6. Tang, S.J.: Research on the spatial layout and site selection of public service facilities based on GIS. D. Central South University (2008)
7. Terefe, A.F.: Application and use of GIS in Small Sanitation Projects in Developing Countries. J. Tampereen Ammatti korkeakoulu (2010)
8. Osumanu, I.K., Kosoe, E.A.: Where do I Answer Natures call An Assessment of Accessibility and Utilisation of Toilet Facilities in Wa Ghana (2013)
9. Feng, J.L., Liu, Y.J., W, G.L., et al.: Analysis of the rationality of urban public toilet spatial layout based on GIS. J. Geospatial Inform. **11**(04), 10–12+185 (2013)
10. Xu, Z.H., Lin, Q., Zheng, W.B., et al.: GIS-based analysis and optimization of public toilet layout in Mawei Majiang area. J. Surveying Mapp. Eng. **24**(07), 24–28 (2015)
11. Xie, J.N., Tan, Y., Liu, Y.F.: Analysis on the rationality of the layout of urban public toilets——taking Yuhu district of Xiangtan city as an example. J. Environ. Sanitation Eng., **19**(06), 4–6 (2011)
12. Gao, Y., Zhang, M.Y., Zhu, Z.P., et al.: Discussion on the spatial layout of public toilets in Qinghe district, Huai'an city. J. Environ. Sanitation Eng. **23**(01), 75–76 (2015)
13. Zhao, Z.L., Cheng, Z.P.: Evaluation of the rationality of urban public toilet spatial layout based on Baidu big data. J. Geogr. Inform. World **27**(02), 49–53 (2020)
14. Shen, A.N., Wang, L.X., Gao, J.: Research on the temporal and spatial pattern of citizen participation in urban governance based on deep learning——taking Suzhou as an example. J. World Geogr. Inform. **27**(02), 44–48 (2020)

15. Zhao, J.S., Zhu, Q.M., Zhou, G.D., et al.: A survey of automatic keyword extraction research. J. Softw. **28**(09), 2431–2449 (2017)

16. Lin, D.C., Tan, M., Liu, K., et al.: accuracy evaluation of representative spatial distribution data set of population——taking Guangdong Province in 2010 as an example. J. Trop. Geogr. **40**(02), 346–356 (2020)

17. Chi, J., Jiao, L.M., Dong, T.: Quantitative identification and visualization of urban functional areas based on POI data. J. Geo-Inform. Surveying Mapp. **41**(02), 68–73 (2016)

18. Yang, J.Y., Shao, D., Wang, Q., et al.: An artificial intelligence method for fine identification of urban land – based on big data of building form and business form. J. City Planning Rev. 1–11 (2021)

19. Zhao, W.F., Li, Q.Q., Li, B.J.: Extraction of hierarchical landmarks using urban POI data. J. Remote Sens. **15**(05), 973–988 (2011)

Extracting Building Contour and Level by Coupling U-net and Single-View High-Resolution Remote Sensing Images

Kaihu Du[1], Boyang Cui[1], Yao Yao[1,2(✉)], Yuyang Cai[1], Yaqian Zhai[1], and Qingfeng Guan[1]

[1] School of Geography and Information Engineering, China University of Geosciences, Wuhan 430074, China
yaoy@cug.edu.cn
[2] Alibaba Group, Hangzhou 311121, China

Abstract. Researches on the urban development and urban planning have an urgent need for building geographic data. Traditional methods of extracting buildings from high-resolution remote sensing images need multi-view images, and have a high cost but a low degree of automation. Thus, these methods are not applied in many fields at large-scale. This study couples U-net and single-view high-resolution remote sensing images to propose a low-cost and simple method for the extraction of the contour and level of the buildings in the remote sensing image. This study adopts the central urban area of Wuhan, Hubei, China as the case study. The results show that the proposed method obtains high accuracies both in identifying building height level (OA = 0.823, Kappa = 0.502) and contour. Compared with the method based on the normalized digital surface model (nDSM), the proposed method obtained a higher overall accuracy of height level extraction increased by 23.4%. The overall quality of building contour extraction is high, and 78.87% of the grids covered by buildings have a building completeness index above 0.4. In addition, we detected and analyzed the changes in buildings in the Nanhu district in the study area based on the proposed method. The results indicated that the height levels of newly added buildings are mainly low and middle levels. The above results have demonstrated the validity of proposed method for extracting buildings contours and level. Moreover, the proposed method can provide scientific supports and reliable help for urban management and renewal.

Keywords: Semantic segmentation · High-resolution remote sensing image · Building height · Contour extraction · Change detection

1 Introduction

As one of the most important features in remote sensing images, buildings are closely related to human production and life [1]. Extracting building height and contours by using remote sensing images can accelerate the update of geographic databases, increase the degree of automated collection, and evaluate the compactness of urban spatial forms,

G. Pan et al. (Eds.): SpatialDI 2021, LNCS 12753, pp. 236–252, 2021.
https://doi.org/10.1007/978-3-030-85462-1_21

which can provide important references for urban land change monitoring and urban three-dimensional modeling [2–5]. Therefore, building height and contours extraction is of great significance to urban planning and development.

Nevertheless, the previous studies are mostly limited by some simple features of buildings, such as the geometry and spectrum, which cannot reflect the relationship between the building and other features. Deep learning semantic segmentation can capture the deep-level features of buildings in remote sensing images, which has shown good performance in extracting building contour and height. Maggiori *et al.* created a remote sensing semantic segmentation model based on Fully Convolutional Networks (FCN) [6]. But when classifying independent pixels, they did not fully consider the relationship with other pixels, causing the lack of spatial consistency and insufficient extraction results. In order to make up for the shortcomings of FCN, Liu *et al.* used the FCN model including Spatial Residual Interception (SRI) to extract buildings [7]. The model uses a volume integral solution to reduce parameters and improve efficiency. It captures and aggregates multi-scale context information by fusing the semantic features layer by layer. Although these semantic segmentation methods have high pixel accuracy, they lack morphological and regular processing of the building extraction.

To tackle these problems, this study proposes a method for extracting building contour and height by coupling U-net and single-view high-resolution remote sensing images. This study uses monocular high-resolution remote sensing images and building vector data to make a semantic segmentation data set, and constructs a U-net semantic segmentation network to train and extract the contours and height levels of the building. We combined the structural differences between buildings and non-buildings in remote sensing images and the differences in materials of different height levels, shadow lengths and building spacing's, and performed multiple classification and contour extraction of building height levels. Moreover, the semantic segmentation results are simplified to obtain the regular building extraction results. Also, we apply the proposed method to building contour extraction, and perform building change detection based on the two-period extracting results.

Our main contributions in this paper can be summarized as follows:

1) This study proposes a set of building extraction methods that couple U-net and single-view high-resolution remote sensing images, which solves the problems of high data threshold, time-consuming and labor-consuming in existing methods.
2) We adopt a building contour regularization method based on grid filling and polygon simplification to solve the problem of irregular edges in the semantic segmentation results.
3) Compared with previous studies on building height estimation using nDSM, the OA and Kappa of building height level classification in this study have increased by 0.234 and 0.261 respectively. Moreover, 78.87% of the grids covered by buildings have the building completeness index above 0.4, which indicates that the overall quality of building extraction is high.

2 Related Studies

In recent years, scholars have used optical remote sensing and Synthetic Aperture Radar (SAR) images to extract the height of buildings. The extraction methods based on optical remote sensing mainly include shadow altimetry measurement, edge detection, and stereo image pairing. The shadow altimetry method is based on the geometric relationship of the sun, satellites, and building imaging with the length of building shadow to explore the height of urban buildings [8, 9]. However, the application of the shadow altimetry method for concentrated buildings with complex structure needs to be improved.

The edge detection method uses isogradient operators to extract the edges of the building itself and its shadow, and then calculates the height of the building based on the geometric relationship [10, 11]. Although this method is simple and easy to operate, it is hard to meet the need for edge enhancement and noise reducing when applied in practical.

The stereo pair method takes a pair of photos of the same area from two orthogonal angles. The stereoscopic observation method and special tools can be used to establish a stereoscopic model of the ground object in the overlapping image part of the image pair. Licciardi *et al.* used the characteristics of image geometric moment invariants to obtain building height from multi-angle high-resolution remote sensing images [12]. Liu *et al.* used remote sensing stereo images to generate a normalized digital surface model (nDSM) to match the building contour to estimate the height of the building [13]. However, it is difficult to obtain multi-angle images data, which is not conducive to large-scale application.

SAR is an active remote sensing that can directly obtain three-dimensional terrain. It is also an all-weather, multi-polarized earth observation method that complements the advantages of optical remote sensing. SAR-based building height extraction methods mainly include SAR stereo image pairing method and SAR and optical remote sensing fusion method. The SAR stereo image pairing method uses orthogonal SAR stereo image pairs to obtain the height of the point with the same name after matching, and combines the DEM data to remove the surface undulations to obtain the actual height of the building [14–16]. Chen *et al.* proposed a building height extraction method integrating the mean shift and AP clustering algorithm with SAR stereo pairs [16]. The fusion method of SAR and optical remote sensing matches the position of the building extracted by optical remote sensing with the three-dimensional data extracted by SAR to obtain a more accurate height of the building [17]. Sportouche *et al.* proposed a "height hypothesis-partitioning generation-criterion optimization" building height estimation method with the fusion of SAR images and optical images [18], and the extraction results are matched with the existing building contour. However, it is difficult and expensive to obtain the multi-view optical remote sensing and high-precision SAR stereo image pair required by such methods, which is why the method cannot be applied for extracting building height in large-scale.

In practice, remote sensing satellites have external azimuth elements when shooting, so most remote sensing images shoot ground objects from an oblique direction [19]. Visual interpreters use the principle of oblique perspective and can judge the approximate contour and height of ground objects based on information such as texture, shadow, and spacing [20]. Computer vision can simulate the visual observation process of the human eye [21, 22]. Semantic segmentation is widely used in remote sensing image pixel classification tasks [23]. The neural network methods (such as FCN) have been applied in building contour extraction in previous studies, which can prove the feasibility of using remote sensing images and deep learning models to extract the contour and height of buildings. Compared with FCN, U-net is a convolutional neural network with a symmetrical structure. It has a contraction path for capturing contextual information and a symmetrical expansion path for precise target positioning and can propagate contextual information to a higher resolution, and obtain building features of different scales, which can effectively solve the problem of insufficient FCN segmentation results, and achieve more accurate segmentation results [24]. Therefore, the U-net semantic segmentation method is adopted to extract buildings from remote sensing images in this study.

3   Methodology

As shown in Fig. 1, the method for extracting building contour and height proposed in this study involves three main parts. First, we use single-view high-resolution remote sensing images and building vector data as the semantic segmentation dataset and adopt U-net neural networks for training. Second, the building contours and height are extracted by the calibrated semantic segmentation model, and then the building contours are regularized based on raster filling and polygon simplification. Finally, we carry out building change detection with an overlay analysis and a raster operation based on the building extraction results.

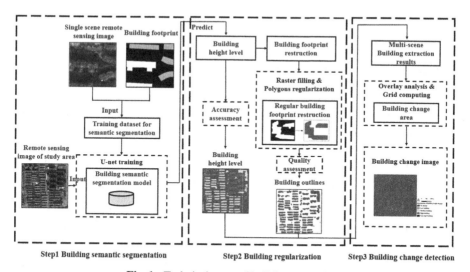

Fig. 1. Technical route of building extraction.

3.1   Building Contour and Height Extraction Based on U-net Semantic Segmentation

In this study, the semantic segmentation method is used to segment the target into several categories. The building height needs to be divided into corresponding categories according to the number of floors. The height of buildings is divided into four height levels, including low-level buildings (<3 floors), middle-level buildings (4–9 floors), high-level buildings (10–32 floors), and very high-level buildings (≥33 floors) [25].

Convolutional neural network U-net is one of the efficient algorithms in deep learning, which combines low-resolution information that provides a basis for building hierarchy classification and the high-resolution information that provides the basis for building contour segmentation and positioning. It can obtain building features of different scales [24], so as to accurately obtain the results of building semantic segmentation. Therefore, this paper uses the U-net semantic segmentation method to extract buildings from remote sensing images.

The U-net built in this study takes high-resolution remote sensing images as input and building classification images as output. The middle part is composed of a contraction path and an expansion path. In the contraction path, each step contains two repeated 3 × 3 unpadded convolutional layers. Both layers are followed by the modified linear unit (ReLU) activation function and a 2 × 2 max pooling operation with stride 2 for downsampling. In each downsampling step, the number of feature channels will be doubled. Every step in the expansive path consists of an upsampling of the feature map. A 2 × 2 convolution kernel is used for the up-convolution operation to reduce the number of feature channels by half. Then the process is cascaded with the corresponding cropped feature map in the contraction path, and two 3 × 3 convolution kernels are used for convolution operation, each followed by a ReLU. At the final layer, a 1 × 1 convolution kernel is used to map each 64-dimensional feature vector to the output layer of the network, corresponding to the category of the target.

3.2 Regularization of Building Contours

Semantic segmentation is the process of classifying each pixel into its category, which is easy to ignore the spatial correlation between adjacent pixels, and the extracted building contours will be rough. Therefore, we adopt a simplification method to regularize the building contours. This method can make the building contours obtained by semantic segmentation more standardized and orthogonal. Moreover, it can also repair the rough boundary phenomenon caused by semantic segmentation to a certain extent [26–28].

This study uses four steps to regularize the building contours: (1) We establish a two-dimensional coordinate system, then use the ray judgment method to find all the grids inside the polygon and fill these grids. (2) We apply morphological expansion and erosion calculations to eliminate the broken points in the grid filling result. (3) We follow the chain coding rule to track the grid in four directions to determine the vector boundary of the target result. (4) The BEND-SIMPLIFY algorithm is used to simplify the polygon, and only the important bends with the curvature radius coefficient r greater than the threshold W are retained to simplify the polygon [29]. The calculation formula of the radius of curvature is shown in formula (1), where y' and y'' represent the first derivative and the second derivative at the bend, respectively. This process retains only small bending points to achieve the effect of polygon simplification (Fig. 2).

$$r = \frac{(1 + y')^{\frac{3}{2}}}{y''} \qquad (1.1)$$

$$y' = \frac{dy}{dx} \tag{1.2}$$

$$y'' = \frac{d^2y}{dx^2} \tag{1.3}$$

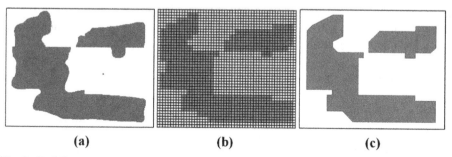

(a)   (b)   (c)

Fig. 2. Building contour regularization process: (a) Original building semantic segmentation results, (b) The result of (a) through morphology and grid filling algorithm, (c) The final result of (b) through the BEND-SIMPLIFY algorithm.

3.3 Accuracy and Quality Evaluation

We use each building as the basic unit to sample one by one and extracts the height at the geometric center of each building. The confusion matrix is constructed according to the extracted height levels and the real levels. We calculate recall rate (Recall), overall accuracy (OA), and Kappa coefficient to quantitatively evaluate the accuracy of the building level extraction results. The calculation method of Kappa is shown in formula (2), where n is the total number of classified samples, t is the number of correctly classified samples, a1, a2… are the actual number of samples for each type, and b1, b2…bn are the number of predicted samples for each type.

$$Kappa = \frac{P_o + P_e}{1 - P_e} \tag{2.1}$$

$$P_o = \frac{t}{n} \tag{2.2}$$

$$P_e = \frac{a1 \times b1 + a2 \times b2 + \dots + ak \times bk}{n \times n} \tag{2.3}$$

In this study, the building contour reconstruction quality evaluation is based on building vector data. We use the common methods of OSM data quality evaluation and the building completeness index (Completeness) to evaluate the reconstruction quality [30, 31]. The completeness index calculation method is shown in formula (3), where **TP** represents the part of the reconstructed building profile that matches the reference building profile, and *FN* represents the part that belongs to the reconstructed building profile but not the reference building profile.

$$\text{Completeness} = \frac{\text{TP}}{\text{TP+FN}} \tag{3}$$

Finally, building completeness is evaluated by dividing a grid of 50 m × 50 m and calculating the quality of the reconstructed building vector. This study has formulated 5 quality levels: building completeness > 0.8, 0.6 ≤ building completeness < 0.8, 0.4 ≤ building completeness < 0.6, 0.2 ≤ building completeness < 0.4, building completeness < 0.2, and building reconstruction Mass distribution map (Fig. 3).

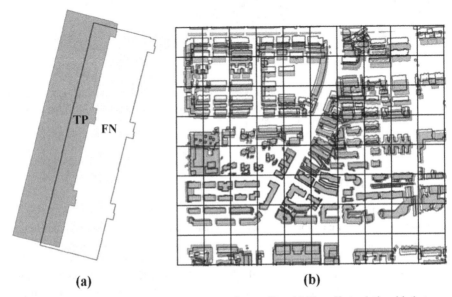

(a)   (b)

Fig. 3. Evaluation method of building reconstruction quality: (a) The offset relationship between the reconstructed building (white) and the real building (grey). FN is the part that belongs to the reconstructed building contour but not the reference building

3.4   Building Change Detection

Building change detection is the process of comparing and analyzing multi-temporal remote sensing images to discover, determine, and obtain building change information [32, 33]. In the process of building change detection, the contour and height of the buildings are the primary factors to be investigated [34]. The basic idea of the building change detection method adopted in this study is to use the two phases of high-resolution remote sensing images at a certain time interval to extract the contour and height of the building according to the remote sensing image building extraction method proposed above. We obtain two sets of building extraction results, and use the methods of overlay analysis and raster operation to obtain both the buildings whose contours and height levels have changed, and the contours of new buildings. Then, we process the contours of the new buildings regularly and add statistics. The area occupied by the building and the level distribution of new buildings are compared with the real building distribution in the remote sensing image, which provides a valuable reference for urban planning and other fields.

4   Results and Analysis

4.1   Study Area and Data

As shown in Fig. 4, the research area is the central urban area of Wuhan, Hubei Province, China. The study data required for the experiment include the high-resolution remote sensing images and the building vector data. The high-resolution remote sensing images are from Google Earth in 2004 and 2019, with a resolution of 0.59 m. We use the contour map of Wuhan buildings in 2018 provided by Baidu Maps as the building vector data, including the number of floors and height of each building. This study creates a building semantic segmentation dataset based on the above data, in which remote sensing images are cropped to obtain training images, and building vectors are converted to raster data and cropped to obtain training labels. Considering the training efficiency, the actual size of the detection target, and the structural characteristics of the U-net, the image and the label need to be cropped to a multiple of an in the range of 200–500 pixels (a is the size of the convolution kernel, n is related to the network depth). Therefore, we choose 224

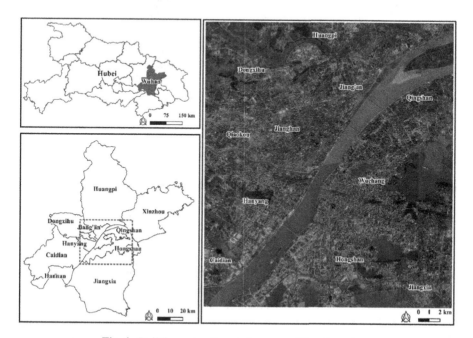

Fig. 4. Building extraction study area and its orientation.

× 224 pixels (between 200–300 pixels, a multiple of) as the size of the training image and the label. The above dataset samples are enhanced by the means of turning left and right and turning up and down, and a training data set containing a total of 16,800 pairs of images and labels is obtained.

4.2 Building Extraction Results and Accuracy Evaluation

The confusion matrix of height level classification in this study is shown in Fig. 5, and the accuracy indicators are shown in Table 1. The Recall of low-level buildings, middle-level buildings, high-level buildings, and very high-level buildings are 0.659, 0.909, 0.684, and 0.556 respectively. The Recall of various categories can reach more than 0.5, the OA of height level classification is 0.823, and the overall Kappa coefficient is 0.502. In the method [7] proposed by using nDSM to estimate height matching with building contours, the classification Recall rates of low-level buildings, middle-level buildings, high-level buildings, and very high-level buildings were 0.840, 0.396, 0.154, and 0.142, respectively. The OA reached 0.589 and the Kappa coefficient was 0.241. In addition to the classification Recall rate of low-level building, which is lower than the result of the method proposed in this study, the classification Recall rates of all the other building height levels are all higher than that of the proposed method. The OA and overall Kappa coefficients of the building height level classification in this study are also compared with Liu et al. The results increase by 0.234 and 0.261 respectively.

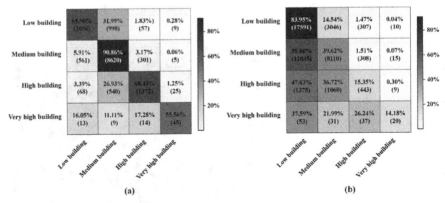

Fig. 5. Confusion matrix of building height level classification: (a) The confusion matrix of this experiment, the numbers in the squares indicate the classification ratio and the number of buildings, (b) The confusion matrix of Liu et al. experiment.

This study uses real building contour as a benchmark to evaluate the quality of the reconstructed building contour. The quality distribution diagram of the reconstructed building contour is shown in Fig. 6. Among the grids covered by buildings, grids with building completeness greater than 0.8 reach 42.02%; grids with building completeness between 0.6 and 0.8 reach 21.37%; grids with building completeness between 0.4 and 0.6 account for 15.47%; and grids with building completeness less than 0.4 account for

Table 1. Comparison of building height level accuracy between the method in this study and Liu et al.

Evaluation indicators	Method of coupling U-net and single-view high-resolution remote sensing image	nDSM method for estimating height and fitting with building contour
Recall rate of low-level building	0.695	0.840
Recall rate of middle-level building	0.909	0.396
Recall rate of high-level building	0.684	0.154
Recall rate of very high-level building	0.556	0.142
OA	0.823	0.589
Kappa	0.502	0.241

21.13%. Moreover, there are 78.87% of the grids have a completeness index above 0.4, which indicates that the building contour reconstruction has a high overall quality.

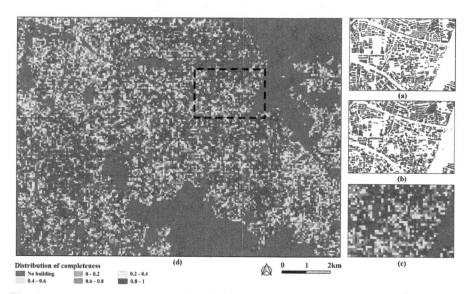

Fig. 6. Building reconstruction quality distribution map: (a) Real building vector, (b) Reconstructed building vector, (c) Building vector reconstruction quality distribution, (d) Vector reconstruction quality distribution of all buildings in the study area.

Figure 7(a) and (b) are the results of the regularization of building contours and the semantic segmentation of buildings, respectively. Figure 7(a1)–(a3) and (b1)–(b3) are the comparison of the local details of remote sensing images and the extracted results. We find that the edges and corners of the regularized buildings are distinct and have the general geometric features of the real building contour, which is more beautiful and regular than the result extracted directly from the semantic segmentation (Fig. 7(a)). Figure 7(b) shows that the boundary of the semantic segmentation of buildings is clear, the predicted building height level is consistent with the real height level in the remote sensing image, and the distribution of mixed building height levels can also be correctly distinguished.

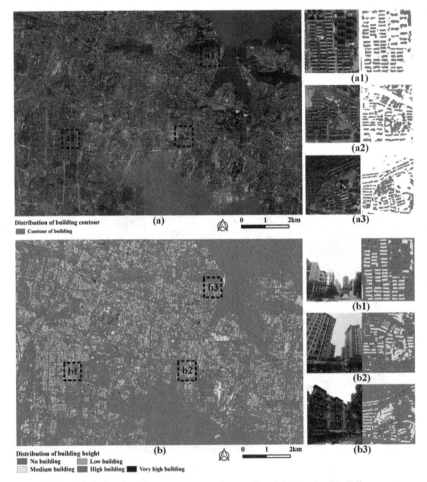

Fig. 7. Building height level and contour extraction results: (a) Result of building contour regularization, (a1)–(a3): Local details; (b) Result of building height level extraction, (b1)–(b3): Local details

4.3 Building Change Detection

This study takes Nanhu Lake area in Wuhan as the experimental area for building change detection, and the buildings within this area in 2004 and 2019 are extracted and compared. The experiment results are shown in Fig. 8 and Fig. 9. The results show that the construction land in this area had a clear trend from north to south during this period, and a piece of farmland in the area was gradually changed to construction land. The new buildings detected in the experiment are mainly concentrated in the original farmland in the south-central part of the area. In addition, there are some new buildings irregularly distributed in various places outside the original farmland, which shows that the area opens up new building land for intensive development, rebuilds and expands buildings on existing plots. The new buildings detected in the experiment occupies an increase to about 3.48% of the total area. Among the new buildings, low-level buildings

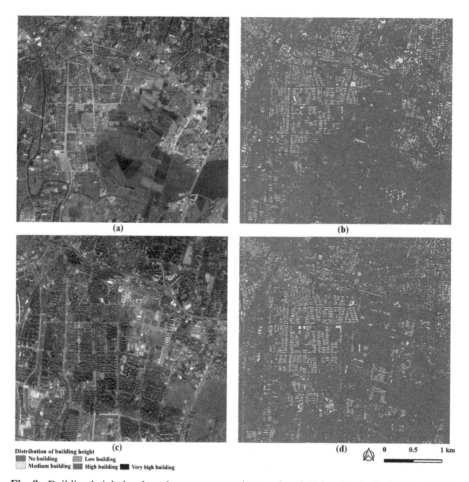

Fig. 8. Building height levels and contour extraction results of Wuhan Nanhu in 2004 and 2019: (a) Remote sensing image in 2004, (b) Building extraction result in 2004, (c) Remote sensing image in 2019, (d) Building extraction result in 2019.

account for about 30.08%, middle-level buildings account for about 43.67%, and high-level buildings account for about 26.25%. The above statistics show that other land-use types in Wuhan Nanhu were transformed into construction land from 2004 to 2019. The distribution of new buildings was mainly low-level and middle-level buildings, mixing with high-level buildings. The distribution of the newly added buildings in the detection results basically coincides with the farmland in the remote sensing image of the area in 2004, indicating that the detection results of the newly added buildings are reasonable.

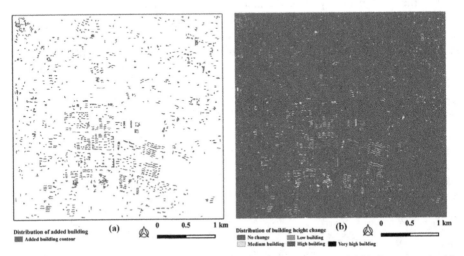

Fig. 9. Distribution of new building contours and height level changes in 2019 compared with 2004: (a) Added building contour, (b) Changes in building height level.

5 Discussion

The accurate extraction of buildings is of great significance to the research of the urban development process and urban planning. However, the traditional high-resolution remote sensing image building extraction research has problems such as high threshold for multi-view remote sensing and SAR data, low automation, and difficulty in large-scale application. In addition, the latest methods such as semantic segmentation also have defects such as irregular building contour extraction and noise points. Therefore, this study designs a set of building contour and height extraction methods that couple U-net and single-view high-resolution remote sensing images. The central urban area of Wuhan, Hubei Province was used as the research area, and the public monocular high-resolution remote sensing images were combined with semantic segmentation and regularization of building contours to enable accurate extraction of building height levels and contours in the research area. Results of the above experiments have proved that the proposed method is both of high accuracy and great efficiency.

This study uses single-view high-resolution remote sensing images for training and experiments, which is free and easily available compared to the expansive professional remote sensing data (e.g., SAR and DSM). The building extraction method proposed in this study has high efficiency. It can extract urban buildings within a range of tens of square kilometers at one time. It takes only about 54s to extract a building in a 5000*5000pixel size remote sensing image. The efficiency is much higher than that of the manually outlined supervised classification method. In summary, this method has the advantages of low cost, high degree of automation, and easy to apply on a large scale.

Since the U-net semantic segmentation model selected in this study takes the image features at multiple scales into account, it has strong ground object classification and analysis capabilities [24]. The building regularization method based on grid filling and polygon simplification make the expression of the building contour more accurate. The combination of the U-net semantic segmentation model with the building regularization method make the extraction results accurately. However, the analytical capabilities of semantic segmentation and building regularization are limited. There is some confusion in the results of building-level classification, and a few grids with completeness indicators less than 0.4 in the building contour quality map. In general, the overall building extraction results are considerable (Fig. 10).

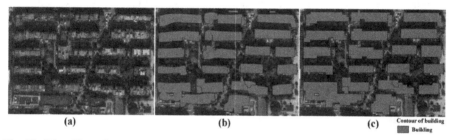

Fig. 10. The effect of building contour regularization based on grid filling and polygon simplification: (a) Original remote sensing image, (b) Convert directly to vector graphics, (c) Convert to vector graphics after regularization.

The proposed method can continuously provide large-scale urban building height level data and contour vector data with the update of high-resolution remote sensing images, thereby speeding up the update frequency of the geographic database. Therefore, it has a high practical value in the field of urban planning and urban land use analysis. The results of the building change detection experiment indicate that a piece of farmland was newly added as construction land from 2004 to 2019, accounting for about 3.48% of the total area. As it is shown in Table 2, the newly added buildings were mainly low-level buildings and middle-level buildings. High-level buildings accounted for only a small percentage. These statistics indicate that during this period, there was an obvious process of concentrated urbanization from suburbs to towns in Nanhu, Wuhan. This conclusion can be verified in the comparison of remote sensing images in 2004 and 2019. In addition, the automatic detection of building changes based on remote sensing images is vital to the sustainable development of urban land use and has broad prospects in the field of disaster assessment [35].

The proposed method has high accuracy and quality, which has been verified to be used in the research of building change detection. However, the small part of the vector

Table 2. Statistics of the proportion of different building height levels.

Detection indicators	New buildings detected by change detection
Proportion of low-level building	30.08%
Proportion of middle-level buildings	43.67%
Proportion of high-level buildings	26.25%

data referenced in the experiment is rough and cannot be matched with the buildings in the remote sensing image, resulting in certain systematic errors in the semantic segmentation dataset. Therefore, the accuracy of this method still has room for improvement. The Recall of such buildings is only 0.556 due to the relatively small proportion of very high-level buildings in urban buildings and insufficient training. In addition, in the building change detection based on building extraction, we have only summarized the overall urban development situation. In-depth internal analysis of specific changes using individual buildings as a unit requires the support of higher extraction accuracy. In future researches, we can consider increasing the number of sampling areas when creating the dataset, and balancing the proportion of samples of buildings at all height levels to train each type of building fully. In addition, we can consider studying the matching algorithm of building vectors, associating buildings with the same name extracted in different periods, and analyzing the changes of buildings in detail to help researches related to land use and urban planning.

6 Conclusion

This study proposes a set of building extraction methods that couple U-net and single-view high-resolution remote sensing images, which solves the problems of high data threshold, time-consuming and labor-consuming in existing methods. We adopt a building contour regularization method based on grid filling and polygon simplification to solve the problem of irregular edges in the semantic segmentation results. Compared with previous studies on building height estimation using nDSM, the OA and Kappa of building height level classification in this study have increased by 0.234 and 0.261 respectively. Moreover, 78.87% of the grids covered by buildings have the building completeness index above 0.4, which indicates that the overall quality of building extraction is high. In addition, we use Wuhan Nanhu as the research area to detect building changes and obtained the distribution of new buildings in the area. The results are consistent with the actual development of the area in the remote sensing image. In conclusion, the building extraction method proposed in this study can help to promote the discovery of building changes in the process of urbanization. The results of this study can also be used in urban land intensity monitoring and urban planning.

References

1. Sirmaek, B., Unsalan, C.: Urban-area and building detection using SIFT keypoints and graph theory. IEEE Trans. Geosci. Remote **47**(4), 1156–1167 (2009)

2. Chen, Y., Tang, L., Yang, X., Bilal, M., Li, Q.: Object-based multi-modal convolution neural networks for building extraction using panchromatic and multispectral imagery. Neurocomputing **386**, 136–146 (2019)
3. Huang, J., Zhang, X., Xin, Q., Sun, Y., Zhang, P.: Automatic building extraction from high-resolution aerial images and LiDAR data using gated residual refinement network. ISPRS J. Photogramm. Remote Sens. **151**, 91–105 (2019)
4. Huang, X., Zhang, L.: A multidirectional and multiscale morphological index for automatic building extraction from multispectral GeoEye-1 imagery. Photogramm. Eng. Remote Sens. **77**(7), 721–732 (2011)
5. Zhao, J., Song, Y., Shi, L., Tang, L.: Study on the compactness assessment model of urban spatial form. Acta Ecol. Sin. **31**(21), 6338–6343 (2011)
6. Maggiori, E., Tarabalka, Y., Charpiat, G., Alliez, P.: Fully convolutional neural networks for remote sensing image classification. In: Geoscience & Remote Sensing Symposium (2016)
7. Liu, C., Huang, X., Zhu, Z., Chen, H., Tang, X., Gong, J.: Automatic extraction of built-up area from ZY3 multi-view satellite imagery: analysis of 45 global cities. Remote Sens. Environ. **226**, 51–73 (2019)
8. Thiel, K.H.: Delimiting the building heights in a city from the shadow in a panchromatic SPOT-image—Part 1. Test of forty-two buildings. Int. J. Remote Sens. **16**, 409–415 (1995)
9. Shao, Y., Taff, G.N., Walsh, S.J.: Shadow detection and building-height estimation using IKONOS data. Int. J. Remote Sens. **32**(22), 6929–6944 (2011)
10. Jahagirdar, A., Patil, M., Pawar, V., Kharat, V.: Comparative study of satellite image edge detection techniques. Int. J. Innov. Res. Comput. Commun. Eng. **3297**(5) (2016)
11. Katartzis, A., Sahli, H.: A Stochastic framework for the identification of building rooftops using a single remote sensing image. IEEE Trans. Geosci. Remote **46**(1), 259–271 (2008)
12. Licciardi, G.A., Villa, A., Mura, M.D., Bruzzone, L., Chanussot, J., Benediktsson, J.A.: Retrieval of the height of buildings from worldview-2 multi- angular imagery using attribute filters and geometric invariant moments. IEEE J. STARS **5**(1), 71–79 (2012)
13. Liu, P., et al.: Building footprint extraction from high-resolution images via spatial residual inception convolutional neural network. Remote Sens. **11**(7), 830 (2019)
14. Soergel, U., Michaelsen, E., Thiele, A., Cadario, E., Thoennessen, U.: Stereo analysis of high-resolution SAR images for building height estimation in cases of orthogonal aspect directions. ISPRS J. Photogramm. **64**(5), 490–500 (2009)
15. Xu, F., Jin, Y.Q.: Automatic reconstruction of building objects from multiaspect meter-resolution SAR images. IEEE Trans. Geosci. Remote Sens. **45**(7), 2336–2353 (2007)
16. Chen, J., Chao, W., Hong, Z., Bo, Z., Fan, W.: Geometrical characteristics based building height extraction from VHR SAR imagery. In: 2017 Progress in Electromagnetics Research Symposium - Fall (PIERS - FALL) (2017)
17. Tupin, F.: Merging of SAR and optical features for 3D reconstruction in a radargrammetric framework. In: IEEE International Geoscience and Remote Sensing Symposium (2004)
18. Sportouche, H., Tupin, F., Denise, L.: Extraction and three-dimensional reconstruction of isolated buildings in urban scenes from high-resolution optical and SAR spaceborne images. IEEE Trans. Geosci. Remote Sens. **49**(10), 3932–3946 (2011)
19. Wierzbicki, D., Krasuski, K.: Determining the elements of exterior orientation in aerial triangulation processing using UAV technology. Komunikacie **22**(1), 15–24 (2020)
20. Van Coillie, F.M.B., Gardin, S., Anseel, F., Duyck, W., Verbeke, L.P., De Wulf, R.R.: Variability of operator performance in remote-sensing image interpretation: the importance of human and external factors. Int. J. Remote Sens. **35**(2), 754–778 (2014)
21. Ahissar, M., Hochstein, S.: The reverse hierarchy theory of visual perceptual learning. Trends Cogn. Sci. **8**(10), 457–464 (2004)
22. Bar, M.: A cortical mechanism for triggering top-down facilitation in visual object recognition. J. Cogn. Neurosci. **15**(4), 600–609 (2003)

23. Long, J., Shelhamer, E., Darrell, T.: Fully convolutional networks for semantic segmentation. IEEE Trans. Pattern Anal. **39**(4), 640–651 (2015)

24. Ronneberger, O., Fischer, P., Brox, T.: U-net: convolutional networks for biomedical image segmentation. In: Navab N., Hornegger J., Wells W., Frangi A. (eds.) Medical Image Computing and Computer-Assisted Intervention – MICCAI 2015. MICCAI 2015. Lecture Notes in Computer Science, vol 9351. Springer, Cham (2015). https://doi.org/10.1007/978-3-319-24574-4_28

25. China, M.O.C.O.: Code for Design of Civil Building (GB50352–2005)

26. Matei, B.C., Sawhney, H.S., Samarasekera, S., Kim, J., Kumar, R.: Building segmentation for densely built urban regions using aerial LIDAR data. In: 2008 IEEE Conference on Computer Vision and Pattern Recognition (2008)

27. Goes, F.D., Cohen-Steiner, D., Alliez, P., Desbrun, M.: An optimal transport approach to robust reconstruction and simplification of 2D shapes. In: Computer Graph Forum (2011)

28. Wang, W., et al.: A Grid Filling Based Rectangular Building Outlines Regularization Method. Geomatics and Information Science of Wuhan University (2018)

29. Wang, Z., Muller, J.C.: Line generalization based on analysis of shape characteristics. Cartogr. Geogr. Inf. Syst. **25**, 3–15 (1998)

30. Fan, H., Zipf, A., Fu, Q., Neis, P.: Quality assessment for building footprints data on OpenStreetMap. Int. J. Geogr. Inf. Sci. **28**(3–4), 700–719 (2014)

31. Girres, J.F., Touya, G.: Quality assessment of the French OpenStreetMap dataset. Trans. GIS **14**(4), 435–459 (2010)

32. Champion, N.: 2D building change detection from high resolution aerial images and correlation digital surface models (2007)

33. Tian, J., Cui, S., Reinartz, P.: Building change detection based on satellite stereo imagery and digital surface models. IEEE Trans. Geosci. Remote **56**, 406–417 (2013)

34. Huang, X., Zhang, L., Zhu, T.: Building change detection from multitemporal high-resolution remotely sensed images based on a morphological building index. IEEE J. Sel. Top. Appl. Earth Obs. Remote Sens. **7**(1), 105–115 (2013)

35. Chen, B., Chen, Z., Deng, L., Duan, Y., Zhou, J.: Building change detection with RGB-D map generated from UAV images. Neurocomputing **208**, 350–364 (2016)

Author Index

Printed in the United States
by Baker & Taylor Publisher Services